PRAISE FOR *SMART*

"Dave Asprey has done it again. In *Smarter Not Harder*, he shows us how to maximize health with the minimum investment of time, energy, and money."—**MARK HYMAN, MD**, *New York Times* bestselling author of *Food: What the Heck Should I Eat?*

"For too long, true health has felt out of reach for too many. Enter this guide for manifesting a radical new level of resilience and recovery. In this deeply researched, practical book, Dave demystifies what it means to be well from the inside out—and he offers a clear path to get there. *Smarter Not Harder* empowers readers with accessible tools to become the agents of their own well-being."—**GABBY BERNSTEIN**, number one *New York Times* bestselling author

"Dave has always been on the cutting edge of what is best for our well-being and longevity. *Smarter Not Harder* is yet another example of his one-of-a-kind vision. Our time is valuable, and the guidance he offers in this book can help us improve our lives in so many ways, and fast. Thank goodness for this road map to better health!" —**DANICA PATRICK**, bestselling author of *Pretty Intense*

"If you want to take ordinary and make it extraordinary, look no further than Dave's latest book, *Smarter Not Harder*. You'll learn how to take full advantage of your body's potential and teach it to work for you, so you can stop sabotaging yourself with fads and restrictive diets that don't work and focus instead on achieving vitality and lasting health."—**VANI HARI**, *New York Times* bestselling author and founder of FoodBabe.com and Truvani

ALSO BY DAVE ASPREY

*Fast This Way: Burn Fat, Heal Inflammation, and Eat Like
the High-Performing Human You Were Meant to Be*

*Super Human: The Bulletproof Plan to Age
Backward and Maybe Even Live Forever*

*Game Changers: What Leaders, Innovators,
and Mavericks Do to Win at Life*

*Head Strong: The Bulletproof Plan to Activate Untapped Brain
Energy to Work Smarter and Think Faster—in Just Two Weeks*

*Bulletproof: The Cookbook: Lose Up to a Pound a Day,
Increase Your Energy, and End Food Cravings for Good*

*The Bulletproof Diet: Lose Up to a Pound a Day,
Reclaim Energy and Focus, Upgrade Your Life*

*The Better Baby Book: How to Have a Healthier,
Smarter, Happier Baby* (with Lana Asprey)

The Human Upgrade™ podcast (formerly *Bulletproof Radio*)

SMARTER
NOT
HARDER

DAVE ASPREY

Thorsons

* * *

Thorsons
An imprint of HarperCollins*Publishers*
1 London Bridge Street
London SE1 9GF

www.harpercollins.co.uk

HarperCollins*Publishers*
Macken House, 39/40 Mayor Street Upper
Dublin 1, D01 C9W8, Ireland

First published by Thorsons 2023

10 9 8 7 6 5 4 3

A catalogue record of this book is available from the British Library

ISBN 978-0-00-862592-4

Health & Fitness

Printed and bound in India by Thomson Press India Ltd

This book is produced from independently certified FSC® paper
to ensure responsible forest management.

Dedicated to every drop of sweat you ever shed that
didn't get you the results you expected.

CONTENTS

INTRODUCTION: BETTER THAN NORMAL 1

SECTION I: RESOURCES FOR LIFE

1. TAP THE POWER OF LAZINESS 11
2. REMOVE YOUR FRICTION 32
3. LOAD UP ON RAW MINERALS...............................59
4. SUPPLEMENT YOUR MeatOS 75
5. CHARGE UP WITH MINERALS 97

SECTION II: TARGETS AND GOALS

6. PICK YOUR TARGET 117
7. HACK TARGET: STRENGTH AND CARDIOVASCULAR FITNESS....129
8. HACK TARGET: ENERGY LEVEL AND METABOLISM 155
9. HACK TARGET: BRAIN AND NEURO-FITNESS 174
10. HACK TARGET: RESILIENCE AND RECOVERY...................194

CONTENTS

SECTION III: ENDLESS IMPROVEMENT

⑪ SPIRITUAL STRENGTH 219

⑫ THE NEXT-LEVEL UPGRADE....................................237

⑬ YOU DO YOU 251

⑭ EVALUATE, PERSONALIZE, REPEAT 263

CONCLUSION: AN OFFERING TO THE WORLD......................272

ACKNOWLEDGMENTS 275

NOTES 277

INDEX 297

SMARTER NOT HARDER

DAVE ASPREY

BETTER THAN NORMAL

Stay curious and grounded.
Be the monk who can meditate at the center of chaos.
Respond with kindness when you see things that feel stupid.

What kind of person do you want to be? Whatever your answer is, this book is designed to help get you there. It is a book about health and fitness, yes, and it is also a book about all the things that you could do if you had more strength and energy. It is ultimately a book about how to be the best version of yourself—you, unleashed.

Everyone talks about empathy and compassion, but you have to combine them with curiosity and equanimity, a profound mental serenity, if you want to hold on to your best self and continue to improve even when things get weird. That combination of qualities is what allowed me to navigate confidently through the past few years of truly weird global disruption. Those qualities, blended together, are also the core principles of living a good and happy life. They are essential in times of crisis, but equally essential when our lives seem superficially dull. And they are, as I'm sure you know, extremely difficult to maintain consistently. Without even realizing it, most of us let them slip away. Sure, we may achieve them from time to time, but then the grind of daily life gets the better of us. We are too exhausted to allow ourselves to be at peace.

Over and over during the pandemic, I've heard people talk about yearning to "get back to normal," and it sounds wrong to me every time. That ambition is just so ridiculously small. All my life, I've worked to improve my physical and mental resilience—to become *better than normal*—and to help other people do it, too. The goal is to reach a higher baseline, make that your new normal, and then reach higher again. A period of crisis is the perfect time to move forward. Why would you want to regress to an old normal?

I want other people to have what I have. To be inspired instead of depleted. To be dangerous instead of afraid. By dangerous, I don't mean that you should do stupid things such as crashing your car or burning down your house. I mean that you can take bold risks, pursue dreams, and be unpredictable because you feel free to act like yourself. It sounds like a paradox, but it's actually an extremely important truth: Being dangerous makes you safe and calm. Being dangerous blows away the sense of impending doom.

Being dangerous also requires a lot of energy and resilience, which is the reason why so many people imagine that a weak "normal" is the best they can hope for. Fortunately, there is another truthful paradox that can help you out. It is technically known as slope-of-the-curve biology, but I prefer to call it what it really is: the *laziness principle*. It is the central idea of this book, and it can transform your life. It boils down to one simple but revolutionary idea:

Laziness can make you strong.

I know, that sounds hard to believe. The reason it's hard to believe is that your body has a secret, one that it doesn't want you to know. Your body is faster than you are. It senses, reacts, and responds to stimuli about a third of a second before your brain even knows what it's up to. Before your rational human brain can apply courage or willpower and hard work, your body has already sabotaged you. It pumps you full of adrenaline that makes small fears feel like big ones. It transmits pain signals to convince you that little tasks are actually going to require a lot of work, which then gives you lots of reasons not to do them.

Why would your body sabotage you like that? Why would Mother Nature create such an unkind system? Because that is the only way things can be in the natural world. Your body is designed to maximize the likelihood that you will survive, have babies, and perpetuate the species. As a result, there are really only two things your body cares about. The first is not dying. The second is being fantastically lazy, so that you use as little energy as possible in the process of not dying.

Your body won't wait for you to decide what to do when a predator is about to eat you. It will start moving you away to safety long before you can make a carefully considered decision about how to react. Your conscious brain is simply not fast enough to respond to threats, which is why it is not responsible for your survival on a second-by-second basis. The autopilot function in your body provides an amazing advantage for keeping you alive over the long haul. It's the reason that humans are still around on this planet. But it comes with some major drawbacks.

During the course of the pandemic, we experienced those drawbacks in full force. We saw how our bodies reacted to the constant onslaught of uncertainty and scary news. At the beginning, our bodies registered the threat and responded with a flood of stress hormones and powerful feelings of anxiety. When the threat didn't go away, our bodies felt as though we were being relentlessly hunted by an unseen predator. They responded predictably—first with stress, then with depression. The reason so many of us were hit so hard is not because we are stupid. It's not because we are weak. It's simply because we are all equipped with an ancient biological system that is trying to keep us alive by making us think we are in charge, when in actuality we aren't.

| WHO IS REALLY IN CONTROL? |

Once you realize that your body makes decisions before your brain does, everything looks different. You can now understand the counterproductive responses that hold you back in life. Better yet, you can start to devise ways to hack your body's systems so that you are in control for real—so that you can make your body do what you want it to do.

The key to taking control is learning how to bend the laziness principle to your advantage. Maintaining laziness is one of the primary biological functions of every cell in your body. No cell wants to use any more energy or resources than the absolute minimum necessary. In response to any stimulus, big or small, all the cells in your body decide how to allocate their energy during that precious one-third of a second before you collect your thoughts. Without fail, they choose the path of less work. That makes sense as a survival strategy: if your cells demanded any extra, unnecessary energy, you might run out of food or you might be too tired to run away from a predator. When in doubt, the safest thing to do is to kick back and relax. If you let them, the cells in your body would be perfectly content to spend their lives living in your grandmother's basement playing video games.

That overwhelming tendency toward laziness is the reason why so many people gained weight during the pandemic.[1] We could all have used the huge amounts of downtime we spent locked in our houses to work out, to meditate, to learn new skills, or to improve ourselves in a hundred different ways, but few took that path. Our bodies didn't want us to do those things; improvement requires energy, and our bodies were in a state of anxiety, focused on using less energy, not more. The laziness principle made binge-watching Netflix seem like a great idea.

The pandemic was different for me. I already knew that my body is designed to feel fear before I get to think about it. I had trained my mind to resist that fear, and I had trained my body to do a better job of distinguishing real threats from small threats. More important, I had begun developing strategies to outwit my body's laziness by working with it rather than trying to fight it.

I want my body to be ready to handle extreme stress and to maintain itself forever. Who doesn't want those things? But my body doesn't *want* to do that. No one's does. The body automatically resists any demands that require a state of higher-than-necessary energy. I've never been good at exercising for an hour a day or even a half hour a day for long periods. You probably know that feeling; the laziness of the body eventually overcomes willpower for most people. Over the years, I've developed a set of biological hacks to guide my body to a more energetic, resilient state, but I always had the feeling that some important element was still missing.

I used the pandemic time to crystallize an idea that's been percolating in my mind for the last twenty years of my biohacking work. There is a more effective way to tell our bodies what we want them to do while still honoring the genetic drive for laziness that keeps all animals alive when things are tight. In fact, *we can embrace laziness as a way to develop more energy.* It's a brilliant shortcut. I didn't want to exercise, but I wanted the results of the exercise. I wanted more results with less effort. By applying the laziness principle the right way, it is possible to do just that.

The standard way people think about exercise is that we have to push the body in sweaty, grunting, painful ways in order to improve. We all know that "grinding it out" on a treadmill is the way to get fit. No pain, no gain! We think about our jobs the exact same way: We fetishize hard work in order to get results. We embrace the "hard work" ethos until we realize it isn't sustainable and we collapse from burnout because our cells think they're being asked for more than they want to give.

What I've come to realize is that there's a way out of this fight. In fact, we can sidestep it entirely, and the laziness principle shows us how. You can trick the very fast but very lazy system that runs your body to get off its lazy ass by approaching exercise (and any kind of work, really) in a totally different way. The standard approach of pushing yourself really hard for as long as possible just sets you up in opposition to your built-in laziness. Instead of working harder, you want to work *smarter.* To hack your laziness, switch tactics and focus on the inputs that laziness understands:

1. How quickly an intense stress comes on
2. How quickly you return to calmness

A high-intensity stress tells your body that it absolutely must respond; laziness is not an option. And a quick return to calm tells your body that it does not need to send you into a state of anxiety, because you are not facing a metabolic energy crisis. In fact, a short, sharp stress followed by an immediate break programs you for less anxiety by training your body to get back to its baseline level more rapidly and efficiently.

If you give your body an intense stress and show your lazy cells that they need to get their asses off the couch, they will. It's just that they will do it a lot more easily and willingly if you let them get right back onto the couch afterward. The speed of a new stress and the speed of your recovery are far more important than the amount of stress. Fast-on and fast-off tell your laziness systems to make you stronger and more adaptable—to give you much better results than you'd achieve if you were doing the slow grind, and to do it a lot more quickly. Creating change in the body becomes vastly easier once you understand this technique.

The laziness principle allows you to reshape your body and mind radically without wasting your time. Along the way, it builds up your power and destroys your stress. It gives you access to the dangerous state I was describing. Powerful people are inherently dangerous. Who knows what they might do? They can handle anything the world throws at them. They can stand up to authority. They can control their emotions so that they are hard to manipulate. They can protect their family and their community. They have the strength to be generous and kind.

Some call that state of energetic flexibility *resilience*, but there are other names for it. In the Buddhist teachings I received in Nepal and Tibet, it is called *equanimity*, which has a richer meaning. Equanimity is the state in which you can remain in control of yourself and be perfectly composed no matter what is happening around you. A lesser state than equanimity is compassion, which itself has value. A lesser state than that is empathy, which has value as well.

| SMARTER, NOT HARDER |

Attaining these elevated states carries a special meaning to me because I grew up with Asperger's syndrome. As a kid, I didn't know how to feel empathy, compassion, or equanimity. I didn't have a clear comprehension of what those things were. In truth, I didn't know how to feel much except the most important survival impulses, fear and laziness. I certainly didn't understand that they were both coming from my body or that they got stronger when my body wasn't healthy. It was a difficult and confusing way to grow up, but that outsider's perspective ultimately helped me understand how to hack the laziness principle.

When I was young, I did what everyone said I should do to get strong: I bought a bike, and I rode it all the time to get in my obligatory sixty minutes per day of steady-state cardio. I slaved away in the gym for hours and suffered through a bland, low-fat, low-calorie diet. I tried to use willpower to overcome my laziness, but my laziness kept getting the best of me as I was left feeling tired and burned-out. Growing up immersed in the fundamental impulses of the body forced me to pay extra attention to the obstacles holding me back. That focus led me to a lifetime of biohacking, to ever deeper investigations of how the body manages energy, and eventually to the book you are reading right now. It took me decades of research and experimentation to reach my better-than-normal state. I'm sharing what I've learned so that you can blow past the obstacles and get there much faster.

A crucial thing to understand is that what is holding you back in life is not an issue of willpower. It is not an issue of weakness or cowardice. It is the *laziness principle* at work, operating outside your conscious experience, and the only way to beat the laziness principle is to embrace it and make it work for you: give your body the right foods and nutrients; apply the right kind of short, intense stimuli; train your body to return quickly to a baseline state. You can train your laziness so that it improves your energy, calms your mind, and expands your possibilities. Take a doctor's advice before embarking on any exercise, dietary or other medical or health treatment.

Once you embrace the smarter-not-harder approach, you unlock the hidden potentials inside you. Then you can become whatever you want to be.

SECTION I

RESOURCES FOR LIFE

1

TAP THE POWER OF LAZINESS

Mother Nature abhors waste. That's why she encoded every cell in your body with programming that instructs it to use as little energy as possible. Think of that programming as a core part of your "meat operating system," or MeatOS. The operating system on your computer does all sorts of stuff that's invisible to you, things that are invisible even to the applications that run on your computer. Likewise, your MeatOS operates undetected in the background. You don't know it's there, but you feel its influence constantly.

The computer analogy makes more sense than you might think. After all, when you click on a screen icon to open your email, you probably don't know which electrons are being sent to what parts of your central processing unit—your "CPU." Maybe you don't even know what a CPU is, and that's fine. The whole point of an operating system is to hide bogglingly complex processes from you. All that matters is the end result: you can run an application that does what you want, without any need to think about how it does it.

Obviously, your fleshy body doesn't run exactly like a chips-and-wires computer, but it must have something like a hidden operating system. Otherwise you wouldn't be able to do things such as take a shot of tequila and break down the alcohol, even if you don't know what your liver is. Your body breathes even without your instruction. Your eyes blink on autopilot when you're not paying attention. The MeatOS works in the background all the time, and it remains entirely invisible to you—at least, until it breaks.

Your MeatOS is not just invisible, it is autonomous. Your brain is not in control of your operating system. Your conscious mind is more like a separate application sitting on top of the many complicated processes that keep your body alive. You never have full awareness of even a tiny percentage of what's happening inside your body. In fact, your limited self-awareness is always out of date. Studies show that your brain doesn't electrically recognize things that happen until about one-third of a second after they happen. Using your primitive MeatOS programming, your body is already taking action[1] about 300 milliseconds before your conscious mind knows what you are doing or why you are doing it.

As a computer hacker, I learned early on how to take over a computer by changing the inputs to its operating system. Later I realized that just as you can hack a computer system, so, too, you can hack your MeatOS. This is *biohacking*, a concept I introduced in 2010.

Hacking is inherently about taking control of a system to make it do what you want it to do—getting the results you want with the least amount of effort. It is the essence of the smarter-not-harder approach. Most of the great early computer innovators were hackers. If something wasn't working the way they wanted it to, they would go into the machine, take it over, and control it. Steve Wozniak and Steve Jobs, the cofounders of Apple Computer, sold an illegal phone-hacking device called a *blue box*[2] four years before they started their company.

An important part of the hacking philosophy is that people should be in charge of the code, not the other way around. When big software companies put out computer code that didn't work the way the hackers wanted it to or that couldn't be customized to their needs and desires, the hackers simply wrote new code of their own. They ended up creating the open-access Linux software, which runs much of the web today; it is probably embedded in the smart TV in your home, too. Hackers set out to get the results they want with the least amount of effort.

Biohackers are continuing this noble (if slightly naughty) tradition, except that in this case, the big software company is Mother Nature. She created a very efficient system, encoded into her MeatOS, that ensures that we unconsciously use as little energy as possible to do the essential things in life: eat, have sex, fight, and form tribes. Mother Nature is relentless in her drive to succeed. By comparison, she makes Microsoft and Google look like Mother Teresa.

The human operating system made us the dominant species on the planet, but it comes with some critical limitations. It isn't designed to make you happy. Or powerful. Or free. Or even calm. All the MeatOS is concerned about is survival and the perpetuation of the species.

That's an unacceptably small, miserable way to live your life. You owe it to yourself, as a good, self-aware, modern human being, to start taking control of your MeatOS. The old programming deep in our bones doesn't work adequately. You face a fundamental choice: either you can be run by your programming, or you can run it. If you want to live fully and freely, there really is only one path. Welcome to being a biohacker.

YOU'RE NOT LAZY—BUT YOUR BODY IS

As soon as you start thinking like a biohacker, you gain new insights into the peculiar, self-defeating things that people do. If something in the body isn't working right, we reflexively search for a short-term fix, because our MeatOS programming makes us lazy. The irony is that the short-term fix is rarely the best one; often it is the hardest, most extreme response. That's why people think they can follow some ridiculous diet for thirty days or kill themselves with maximum-suck workouts at the gym and find some magical solution to their problems.

Biohackers are more patient, methodical, and efficient. We are willing to try different techniques until we figure out what really works. We are open to unusual solutions, we put everything to the test, we run experiments on ourselves, and then a few of us share our best results with the world. Although biohacking often makes use of the latest science and technology, it is anything but artificial. It is actually a return to nature, because it honors our lazy programming rather than fighting against it. Biohackers reject modern synthetic foods and wasteful exercise techniques that are not—and never have been—compatible with the natural operation of the MeatOS.

Every hacker needs a way in, a method of gaining access to an operating system in order to manipulate it. For your body, the laziness principle is the way in. It gives you root access to the MeatOS and allows you to make extraordinarily powerful adjustments to the way your body functions, so that you can get a higher return on all of your investments of time and energy. You are so beautifully, elegantly lazy on the inside that your body knows exactly which systems to shut down, and in what order, if you don't have enough energy. You can use those systems to your advantage, retraining them to make your body allocate energy where you want it. Do you want to live longer than nature designed you to live? You can do that. Do you want to have more energy? You can do that, too. Smarter? Check. Faster? Check. At peace? Check.

Taking control of your biology to realize more of your potential is not a new idea. People have been seeking ways to do this for thousands of years, using tools including food, drugs, exercise, religious ceremonies, and meditation. It's just that now we can measure what works very quickly and change what we do until our biology responds in the way we want it to. We can no longer convince ourselves that something is working just because we want to believe it does.

With the latest biohacking techniques, I can help you master your MeatOS and improve yourself effectively and measurably—and do it using far less time and effort than you would spend on standard techniques such as brutal gym workouts. When your body has the right raw materials available, you can send it the right signals in order to get all the benefits (and much more) of a full-on, exercise-obsessed lifestyle in 90 percent less time. If you're spending forty-five minutes a day working out, that adds up to more than twenty thousand hours over a lifetime—like ten years working a full-time job. I want you to get that time back. And if, like most of us, you don't have that kind of time to sweat and grunt in the gym, I want you to get the energy and happiness you deserve without ripping up the rest of your life. Use the extra time to meditate, to have fun, to build something great, to play with your kids—or even to make some.

I know that it's possible to hack your biology, because I've seen it time and again. I grew up as an overweight kid dealing with intense immune issues, arthritis, and brain fog. After I started biohacking, I launched what is now a big company while working a full-time job and dealing with health problems that had plagued me for years. Today I am a dad, a podcaster, and an entrepreneur who advises dozens of startups and manages seven different companies of my own. I have 11 percent body fat, and I'm never hungry. Using my brain is so effortless that it just feels good. When I want more muscle or endurance, I know how to make it happen. If a fat computer hacker like me can end up shirtless in *Men's Health*, you can do it, too. And I can show you how.

My latest version of biohacking is more powerful than ever, because it incorporates new insights into how the laziness principle works. By putting the correct resources and signals into your body, you can think more clearly. You can make your muscles and nerves operate more efficiently. And you can ditch the standard gym workout in favor of an upgraded set of routines that are more effective but drastically less draining.

There's a powerful reason why gearing up to go to the gym can seem as though you're getting ready to climb Mount Everest. That's your MeatOS using the laziness principle to make you procrastinate. How many times have you gone out to eat fast food because your lazy operating system didn't want you to cook or do dishes? If you could have had a delicious, healthy home-cooked meal with the same amount of work (or less) as eating out, you'd gladly have done it. But your MeatOS convinces you that going out is the lowest-energy thing to do, so you tell yourself stories about why going out to eat is the right choice. Then, after you eat the last French fry and check your bank account balance, you wonder why you keep doing it. It's because your body is lazy and will do anything possible to avoid work unless the work is worth it. But when you honor the fact that your body is lazy (even if *you* aren't), you'll be happy to do a few simple things that give you a lot more results.

TAMING THE LAZY BEAST WITHIN

It's time to get to know yourself better and explore the inner workings of your biological laziness. Everything alive has evolved to automatically maintain a stable inner environment when external factors change, with no brain action required. The name of this equilibrium state is *homeostasis*. Without biohacking, you leave your MeatOS completely in control of homeostasis. You don't consciously manipulate your heart rate, blood glucose, or brain waves. They regulate themselves based on what they perceive.

When things are in a state of balance, all is good. But when something pushes your body out of homeostasis—whether by illness, stress, injury, nutrient deficiency, or exercise—your body will do one of three things.

Die. In the worst case, your body will fail because it wasn't resilient enough and you will be either dead or disabled. (Your body's job is to be resilient, so that's unlikely.)

Change. In the best case, your body will change and adapt to better handle the situation. Sadly, this usually takes a lot of energy, which your MeatOS doesn't want to spend.

Avoid. In the most likely case, your body will convince you not to repeat the thing that pushed you out of balance, because it wastes energy and might even be dangerous. This is why most people stop exercising over time.

Your MeatOS may not be smart, but it is inescapable. It is built into every organism made of meat. It has been honed by millions of years of evolution, so it is extremely good at keeping you alive—and keeping you lazy. You're stuck with the MeatOS if you're alive, so it's up to you to apply the smarts that it doesn't have.

Remember that there is nothing evil about the laziness principle. In many ways, it is quite marvelous. It is what has driven every major human innovation. Think about it: Why did we invent fire? Because it was easier to stay warm that way than to shiver to produce body heat. Why did we invent spears? Because they are more efficient weapons than clubs are. We invented washing machines because people were wasting two hours a day scrubbing clothes on washboards. We like to go out to eat because food preparation and kitchen cleanup are a time suck. The single greatest driver of human progress is not love; it is laziness. (Fortunately, love is a close second and is also built in to your MeatOS.)

Accepting that your biological laziness is your best and most motivating attribute can seem terrifying at first. It goes directly against the hard-work ethic that most of us have had drilled into our skulls since childhood. It took me a long time to embrace the concept. But here's the thing: laziness and achievement are not at odds. The laziness in your MeatOS can help you win because it makes you efficient. Biohackers are proudly the laziest people on the planet.

I first became aware of the laziness principle many years ago, long before I fully understood where it came from. At the time I had just started one of my first jobs, working at an obscure type of company called a food brokerage. These companies are part of our byzantine food distribution network, performing the important task of making sure a certain brand of canned tuna gets the best shelf space in your local store, things like that. I worked in the IT department, in charge of the computers that tracked food shipments and deliveries.

My work was eye-wateringly boring, so my laziness principle automatically kicked in. I realized that I could automate a lot of my tasks. Instead of managing each computer separately, I could write software that would automate them all. It let me manage twenty computers with time left over even though management expected a normal IT guy to manage at most ten computers. If I could automate six hours of my daily work, that gave me six hours a day to do what I wanted to do—which in this case, because I was a geek, was learning more about the latest computer tech so I could be even more lazy. Or have more control; they're opposite sides of the same coin.

Then I moved on to a job at the first internet data center company, back in the early days of a newfangled thing called the World Wide Web. The idea was to help companies (including Google and TheFacebook when they had only a dozen employees) add thousands of computers to their services as fast as possible. My job was to hire people to run all those computers. No matter what I did, I couldn't hire enough people fast enough (eventually the group I cofounded had a thousand employees), and they kept making mistakes.

Laziness stepped in, and it became obvious that we would have to create technology so one person could manage a million computers. That way, we computer hackers could kick back, drink some more coffee, and figure out how to make the systems manage themselves. Yes, it was pure laziness (plus money) that led a few internet geniuses at the time, including Marc Andreessen (who created the first web browser) and the folks I was working with, to invent what we now call cloud computing. Behind the desk where I write, I have a poster on the wall from the launch of the first-ever cloud computing service. It reminds me every day to be grateful for laziness.

I don't want to do more work than necessary. Neither do you.

It's easy to embrace the idea that the laziness mindset is great for tech innovation. Everyone loves efficiency and productivity, right? But somehow, we are embarrassed about applying the same mindset to our own bodies or well-being. I'm giving you permission to call BS on that. The idea that hard work is the only honorable work is puritanical nonsense, created hundreds of years ago to trick your MeatOS into being willing to do farm or factory labor to benefit somebody else without revolting.

We all believe that change is hard because we hear that message all the time. Society repeats it to you. The laziness in your MeatOS repeats it to you as well. And your daily experience seems to confirm that it's true. Running is hard. Lifting rocks is hard. Fasting is hard. Meditating is hard. Even if *you* want to do those things, your operating system doesn't want to waste energy on them. After all, if it lets you waste energy doing that stuff, it will have to spend even more energy to change and adapt. To get lean. To get smarter. To be calm. It's simply less work to stay fat, slow, tired, and tweaked.

The number one hurdle to becoming a biohacker is allowing yourself to embrace a radical truth: that there is a simpler way to tell your body to change, one that doesn't take as much work. You can change how your body and your mind work—without a lot of grief—by sending your MeatOS the right signals. You just have to convince it that it must abruptly leave its effortless state of being.

THE POWER OF THE SPIKE

Imagine a signal that convinces your automated system that it has to change or it might die, without actually causing that much real danger. Imagine, further, making the signal so brief and simple to create that it doesn't take you much effort, not even enough to trigger your innate laziness. Now you don't have a reason *not* to do it, and your system will still change as fast as humanly possible, which is a lot faster than you may think.

Pay attention to the way that the signal increases and decreases over time. A quick, sharp, intense challenge to your current (tired, overweight, stressed) state helps you achieve dramatic, rapid changes—and it's those kinds of changes that have the biggest impact on your resilience and well-being—unless, of course, you send the wrong signal or one that's too strong. That can cause your body to lose its natural state of homeostasis so you lose the benefit of the signal. Think of how you feel when you overtrain or pull an all-nighter.

Here's a graph of a typical cardio workout, which looks suspiciously like all the forms of cardio exercise you've ever seen (jogging, spin class, stair stepper, and so on). It shows how much energy you expend over time:

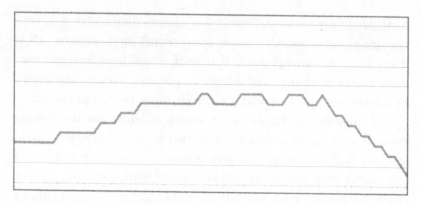

An hour of grinding is a lot of work, so of course your body hates it and your innate laziness will try to trick you into not doing it and eating a doughnut instead. Sometimes you'll use willpower (which takes more energy) and make yourself do it anyway. But it's a grind until you're done, when you might feel more energized—or you may just be tired.

Now let's look at a smarter approach. Try something new that you'll learn about in this book, and your workout will look like this instead:

You've saved a lot of time, and you pushed yourself harder, but you did less work overall. From your body's MeatOS perspective, it will have to change itself to adapt because you sent a signal that it has to be prepared for more intense levels of performance. But from your perspective, it was easier because you didn't have to push your lazy body for as long. You got the same benefits as with the sixty-minute grind workout, but it sucked for only fifteen minutes.

Can you do even better? Sure, you can. As a biohacker, you have given yourself permission to double down on your laziness. After all, if you can free up some time or gain more energy, you might be able to do something meaningful.

So you harness your lazy drive and use it to figure out how to get the same (or better) results with even less suck. You harness all kinds of data and test different outcomes. Or maybe, being lazy, you just peek over a researcher's shoulder during the final exam and copy his answers. Either way, the end result looks like this:

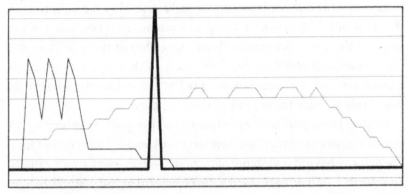

Now you're down to less than one minute total of suck, which most people can handle. But your body is freaking out, being tricked into believing that in order to survive, it's going to have to create a strong, resilient system capable of handling that single spike of exertion.

This principle will set you free.

It doesn't apply just to exercise; it applies to almost all of the things we now know can radically reshape how your body looks, upgrade how your mind works, and create a kind of resilience that seems almost superhuman. It will make you realize that you can do anything. It will make you inherently dangerous because you will no longer be under the control of your lazy nervous system.

In technical terms, I call this spike style of activity *slope-of-the-curve biology*, because it creates a steep slope (rapid increase, rapid decrease) in the signal you are applying. To put it simply, your body doesn't care how much time you do something hard; it cares about how quickly you do something hard, how hard it is, and how quickly it returns to baseline. I've founded two companies, Upgrade Labs and 40 Years of Zen, to figure out exactly what signal to use, how strong to make it, how quickly to turn it on, and how quickly to turn it off in order to make your body and mind rapidly get better at doing anything. The answers won't be exactly the same for everyone (an obese sixty-year-old is a lot different from an athletic eighteen-year-old), but everyone can use the techniques I've discovered in these labs, often without using any technology at all.

You will discover that you don't need to waste an hour sweating in a group fitness class. If you do the right actions at the right time in the right order, focusing on rapid and intense changes, you can get slope-of-the-curve benefits in about one-eighth of the time. You can also design a targeted recovery program so that you will return to a calm, relaxed state much more quickly and efficiently, which tells your body to improve instead of freaking out.

In this book you will also learn that the goal of *returning to baseline* applies to every intervention that you do. The secret of feeling calm is having a system that doesn't waste energy and rapidly returns to baseline normal, whatever normal is for you. Then you can adjust your normal to be better and better. If you can go to full power and then return to a relaxed state in almost no time, you will become stronger, calmer, and clearer more quickly than you ever imagined was possible.

Traditional workouts, diets, and even meditation are designed to get you to 70 percent of full power and keep you there for a long time. What an enormous waste of effort! A long-duration, 70 percent workout sucks up all of your energy without ever taking your body to the sharp peak where the real improvement occurs. No wonder so few people exercise! The way we work out today is inherently wasteful, and it goes against our nature. Since our nature runs our body before our mind can even think about whether we want to exercise, we just shrug, let the laziness principle do its thing, and eat the potato chips while feeling ashamed that we skipped the gym. Again.

Don't believe me? Mother Nature offers plenty of evidence, if you know what you are looking for. Watch a nature video of a deer

escaping from a mountain lion or any animal escaping from a predator. The prey sprints away at top speed, at full exertion. If the
prey gets away, it just stops. Its body shakes from nose to tail as
the animal lets go of stress hormones, and then it resets itself back
to baseline and sniffs out some fresh grass to eat. That is powerful,
beautiful natural laziness in action, without a drop of wasted energy. You know what deer *don't* do? They don't hunker down for a
ninety-minute cardio workout on the elliptical to stay lean.

You want to be like that deer, relaxed but hard to kill. You want
to be so powerful that when a challenge comes your way, every system in your body can come alive with incredible energy, focus, and
courage. And after you vanquish the threat, your automated system
will quickly return to normal. You can simulate the rapid-on, rapid-
off response of a deer escaping from a predator by giving yourself a
rapid-on, rapid-off workout. It's a totally different technique from
the standard gym or jogging routines. You can train your brain to
be similarly resilient by using sharp, targeted signals: light, sound,
guided breathing.

Look, life throws us curveballs all the time. You may not be
chased by a tiger, but you might be in a car accident or get sick. You
might suddenly lose your job or a loved one. You might wake up one
morning to find yourself living in the midst of a global pandemic.
Your MeatOS does a terrible job of distinguishing a bad breakup
from a tiger and will turn on your stress response either way. If you
don't make enough energy because your lazy system got its way, the
stress will knock you out of equilibrium, out of homeostasis. Your
system will degrade. Your body will become even worse at making
energy. You will become anxiety prone, depressed, weak, indecisive.
You will feel helpless.

On the other hand, if you've used the knowledge in this book
to your advantage, you will have enough energy left over to do the
right things: To think. To return to baseline. If your MeatOS is
programmed to throttle on and off quickly and easily, your system
can take a big hit without losing its stability. Your system can also
handle a lot more low-level, daily stressors when it is trained and
ready to handle big changes. You will be able to come home after

a long, particularly hard day at work, shake it off like the deer did, and still have enough energy to smile at your spouse and your kids. Or maybe you will go party with your friends. The choices and possibilities are nearly endless when you are operating at full power.

THE PRINCIPLE OF ENERGY

Everything you think and do depends on energy, so if you want to be an effective biohacker you need to track that energy back to its source: the mitochondria,[3] the pill-shaped miniorgans (also known as organelles) that live inside your cells. Science textbooks typically refer to the mitochondria as the "powerhouses of the cell," because they produce energy in the form of an amazing molecule called adenosine triphosphate, or ATP, which your body then uses to fuel your day-to-day activities. The assembly and destruction of ATP molecules within the mitochondria power your entire biology: everything you do, feel, think, and dream.

Although the mitochondria have power, they do not have control. Even they have to answer to the MeatOS, which tells them when to activate and how hard to work. Under orders from the MeatOS, the mitochondria become the source of that sacred laziness that drives us to create new and better ways to do less work. Each mitochondrion is an environmental sensor, a tiny computational node that makes decisions, a manufacturing plant, *and* a power plant. The mitochondria function as a distributed battery, too, allowing your body to maintain a charge equivalent to about that of an AA battery at all times. They can even talk to one another, participating in an organized voting scheme called *quorum sensing* so they can decide what's really happening to your body and respond in unison.

In short, your body's energy system acts as its own independent intelligence unit embedded within your cells. That system is very fast, but also very lazy and reactionary. When you own it, you are very fast. When it owns you, you are lazy and reactionary.

Everything about your body's energy system—the mitochondria and the MeatOS that regulates them—is ripe for hacking. In fact, I've never in my life seen a system so ripe for hacking! I've designed and built computerized systems with tens of thousands of nodes in thousands of locations across the globe and taught postgraduate professional classes how to make them faster, stronger, and more secure. I know how to spot inefficiencies and how to improve them. Surely we can consciously train and shape the automated intelligent systems inside our bodies so they will stop doing the things that make us feel weak and tired.

And so begins the next level of the hack: figuring out the proper feeding and training of your mitochondria. They form a complex organized system, acting almost like a separate colony of organisms living inside you. Each one has its own DNA, distinct from the DNA that contains your overall genetic code. On the outside, the mitochondria have a double membrane made of fatty molecules that protects them and regulates what chemicals enter and leave. On the inside, they contain an elaborate set of molecular machinery that depends on having exactly the right chemicals on hand. In total, your body contains about 10 quadrillion mitochondria, and when you exert yourself, those little things produce about one pound of ATP every minute.

Some of your cells contain more mitochondria than others. The energy-hungry cells in your brain, eyes, and heart, in particular, are teeming with these little guys. Muscle cells carry a medium-sized mitochondrial crew. Your cells build more mitochondria through a process called *mitochondrial biogenesis*. The more mitochondria you have and the more efficiently they operate, the better you feel. When you have plenty of mitochondria working at full capacity, you have energy, strength, resilience, and clarity. On the flip side, if your mitochondria are declining in number or not operating efficiently, your MeatOS will feel anxious, and you will, too. You will feel sluggish and foggy; you will gain weight and try not to think about the fact that you are not at your best.

A remarkable, paradoxical insight of the laziness principle is you can never feel truly calm if your MeatOS is weak. It takes a high energy capacity, with a strong energy supply at the ready, to remain truly relaxed in a calm, baseline state. In other words, you must have the power to be dangerous in order to be at peace. Weak cells and weak people know they are never truly safe, because any slight insult could trigger the loss of equilibrium. Strong cells and strong people have the energy to stay young, active, and dynamic. They can follow their desires to do things that matter in the world.

Energy is the most precious resource in life. That's why biohacking begins by finding the tools that will work best for you to gain control of your energy system while honoring your inherent biological laziness principle. All other improvements grow from there.

If you have time and energy, you can make money.

If you have money, you can buy free time.

If you have no energy, you will spend all of your time sleeping and all your money getting your energy back.

Most of us experience a "no-energy" state as we age because our mitochondria become fewer and weaker over time unless we hack them. As a young man, I got caught in that no-energy state, and I spent all of my time and more than $1 million trying to get my energy back. I was stupid and wasteful. I won't make that mistake again.

Most of us are stuck where I was: Low energy. Tons of stress. Overwhelmed. Brain fog. Sleepless nights. In fact, today more than 73 percent of American adults are overweight or obese, and 42 percent are obese. Only sixty years ago, less than one lifetime, just 31.5 percent were overweight and 13.4 percent were obese.[4] Obesity is the number one sign that your mitochondrial MeatOS is not making the energy you need to transform yourself.

It is a fool's errand to spend vast amounts of energy trying to make your body do things it naturally doesn't want to do. If you want to relax, you need enough energy to feel safe. If you want to be stronger, you need enough energy to build muscle. If you want to be smarter, you need to turn up the energy in your head. If you want to be a master of your craft, you need enough energy to fully immerse yourself in it. If you want to be a kinder, nicer human being, you need to produce enough energy to be far, far more than just enough.

To reach your better-than-normal state, you must have both the energy and the knowledge to bend your MeatOS and your biology to do your bidding. After all, if you create huge amounts of energy but don't direct it, your MeatOS will happily apply that energy to watching Netflix and eating takeout. So let's get smarter.

SIX STEPS TO ENERGY SUCCESS

Everything you do sends a signal to your body about how you want it to perform, but most of us make choices without having a clear understanding of their consequences. We eat foods that hold us back. We exercise in ways that work against our inner laziness. We are unaware of simple inputs we could apply that could have huge biological advantages.

Fundamentally, hacking the laziness principle is a six-step process. Reprogramming your MeatOS so that you have the power to be your truest, fullest self begins with providing the raw materials to make energy. Then come convincing your body to make more energy (without triggering the lazy factor) and putting that energy to work on the specific systems you want to improve: your metabolism, your cardiovascular performance, your stress response, your strength, your brain. There is convincing new science in each of these areas that will radically reduce the amount of struggling necessary to get there.

Step 1: Remove friction.

Your body automatically conserves energy, so if your cells can't make it very well, everything in your life will feel hard. The first step is to figure out what you're doing that's breaking your energy production. If you're feeling stressed, sick, weak, or foggy, even the easy hacks are going to feel insurmountable. In chapter 2, I'll provide a detailed list of obstacles (especially dietary ones) to remove from your life so you can quickly develop more energy. This is the low-hanging fruit of the book. It is far easier to stop doing things that make you weak than it is to start doing things that make you strong.

Step 2: Load up on raw materials.

In superhero movies, you always see the main character running around, fighting bad guys, and performing over-the-top stunts, never stopping to eat and drink. Here's a tip: if you were one of those superheroes, you'd be consuming 50,000 calories a day to maintain all that power (unless you could just suck pure electricity, maybe). It is far easier to be a fully realized, fully energized human being. To do that, you need enough calories; after all, calories measure energy, and you want *more* energy. But even with enough food, your MeatOS can't direct all those little mitochondrial power plants to do their magic if your system is short on the right elements and cofactors.

There are a few foundational resources your body needs to make full use of its abilities, and most people are deficient in them. If you focus on getting the right fats, fat-soluble vitamins, and big and microminerals, plus enough high-quality protein, you will unlock new levels of energy. That, in turn, will make you feel less lazy or more motivated. Whatever you want to call it, it's an upgrade. I'm not going to tell you to take a fistful of expensive supplements. But I will explain why, no matter how well you eat, in the modern world you cannot get all of the resources you need from your food alone. You'll learn to identify the vital, must-have supplements, and you'll read about optional ones that have profound effects.

Step 3: Pick a target area and track it.

Deciding what your goals are is a process in itself. There are five big areas of biological performance you can choose to control and improve. They are: strength, cardiovascular fitness, energy and metabolism, brain function, and stress resistance. The good news is that when you focus on improving one, the others get better, too. You need to pick a primary target, though, because your strategic laziness function will get overwhelmed if you decide to upgrade everything all at the same time. Don't worry: you'll have plenty of energy left to upgrade the rest of them after you do the first one.

Step 4: Send the right signals.

At every level, you can use the slope-of-the-curve principle to change the signals you send your body. It will respond by improving quickly and dramatically. For every area you want to work on, you'll have a choice of free techniques you can use at home, low-cost portable technologies, and cutting-edge equipment you'll most likely have to seek out from a professional. Regardless, there are always simple, accessible ways to get started.

Many of the most powerful and precise ways of getting a stronger signal into your body are ones you won't find in a traditional gym: light, sound, vibration, lymphatic drainage, electrical stimulus, heat, and cold—even cryotherapy. You don't need to do any of these, but I describe them because they demonstrate what is possible today. I'm working quickly to expand my chain of Upgrade Labs to make them accessible in many major cities. When I was in my twenties, I spent nearly two years getting strong enough to max out on the strength machines at my gym. I know now that I could have done more good for my body in a tiny fraction of the time, with less chance of injury, less soreness, and less exhaustion. I don't want anyone else to repeat my mistakes.

Step 5: Recover like a boss.

It's exciting to push yourself to new heights and improve. It's boring to recover as fully and intensely as you stimulate yourself. That mismatch happens because you get exciting and beneficial stress hormones such as adrenaline and endorphins from pushing, but you don't get them from recovery. Nonetheless, slope-of-the-curve biology teaches us that if our bodies recover fully and quickly, they will devote all the extra energy to growing stronger.

If your body recovers less quickly, you will evolve less quickly. If you want the maximum biohacking benefits, you need to dedicate yourself to the recovery process. It's as simple as that. Recovery signals and technologies also pull double duty as stress management techniques, so if your primary focus is stress reduction, you may want to start here.

Step 6: Evaluate, personalize, repeat.

For every hack, I will walk you through the qualitative and quantitative ways to evaluate what is and isn't working for you. You will learn how to look at your tracking data, refine your hacks, and go back to step 3 so that you can keep improving.

You'll learn about three types of tracking: real time (with a fitness tracker or continuous glucose monitor), daily (such as sleep or stress metrics), and occasional (lab tests or imaging). Basic monitoring is either free or close enough to it to get you started regardless of your situation. You'll be able to tell if your energy levels are higher, your vision is clearer, or your mind is sharper. But you will also want to get hard numbers: You recovered better last night. Your blood sugar was great at your last meal. Your bone density increased. Your inflammation markers went down.

Biohacking is not based on blind trust. You can, and should, measure it objectively for yourself. You should also adjust it to your particular goals, situation, lifestyle, and physical state. Just because a hack works well for some portion of the population doesn't mean it will work for everyone. In the age of advanced self-tracking, you can iterate, refine, and create a routine that's right for you.

| ENERGY FOR A BETTER WORLD |

At the end of the day, biohacking is my business, but it's not my motivation. My deeper purpose is this: I want to save you from the terror of not having control, at least enough control, of your own biology. I know that when you develop more energy, it fuels everything you want to do in your life, including just being a better person.

Right now, humanity is at a crossroads. The nearly 8 billion people on Earth act as a single cooperative organism with an enormous number of decision-making nodes called individual humans. Each of those people, in turn, is an organism with quadrillions of its own little decision-making nodes running the MeatOS. Collectively, that's a staggering amount of power and complexity, a system unlike anything that has ever before existed on this planet. Every morning I read the headlines and it strikes me how much capacity we have to operate in kindness and generosity—or to immerse ourselves in hatred and selfishness.

We have to fix our own operating system, our biology. Otherwise we're on a path to kill one another and destroy the planet. It's time for a human upgrade, because we're all a part of the big and precarious human organism. If we are going to heal as a society, we must begin by healing ourselves, and that begins by creating full-power biology that honors our inner workings instead of ignoring them. We may be motivated by laziness, but we are not lazy; we are powerful. We are wired to handle this, and so much more.

2

REMOVE YOUR FRICTION

To live, you need the right resources to generate energy. To live *well*, you require a lot of those resources, in the right mix—and you almost certainly are not getting them. Your conscious mind might want you to become stronger and more resilient, but your inner operating system has other ideas. If it senses a shortage of raw materials, it goes into self-preservation mode: it generates friction, slows you down, and keeps you from going where you want to go, no matter how hard you work.

Every second, your MeatOS automatically makes tens of thousands of microdecisions to ensure that your body works well enough to keep your brain alive. Because you can't directly observe this biological programming, it is easy to screw up without knowing it. If your diet is deficient in the raw materials of life, it's not as though you are starving. You don't drop dead. For a while, you probably don't feel anything at all. Yet at the microlevel, your body is making little adjustments. If you're short in even one trace mineral, your MeatOS elegantly identifies the least important function that mineral supports and switches it off.

Your brain still functions. You are still breathing. But one small part of your biology is weakened. Your body's stress system becomes microscopically stressed. Your long-term repair systems take a tiny hit, because they are the least important systems for keeping you alive right at this moment. You are less resilient, and your body has less energy for you to use. Sometimes your MeatOS will leave a biological system switched off even after you start getting the necessary resources, because it is relentlessly guided by the laziness principle. Unless your MeatOS can be convinced that it has the energy and the raw materials to turn that system back on, it won't do it.

Your current diet is probably not giving your body the nutrients it needs to optimize its energy supply. Even worse, you are probably filling your body with antinutrients—natural and man-made chemicals that prevent you from absorbing the nutrients you do eat and that even suck minerals out of your tissues. No meditation or workout is going to overcome that shortfall. Fancy supplements won't help if your body is depleted in the basics. A lack of raw materials is one of the biggest blockers of your energy, longevity, and performance.

Let's get smarter, fix the problem, and remove the friction so that you can move ahead in your life.

WHAT YOU NEED THAT YOU AREN'T GETTING

Everyone knows that minerals are good for you, but few people get enough of them because they simply aren't the sexiest or best-marketed supplements. Herbs and nootropics get all the top billing, but common minerals are glossed over.

I run my own farm, which has really opened my eyes to the importance of minerals. As I write this, I'm looking out over a couple acres of mixed fruit and vegetables and spices, along with three cows, twenty-five sheep, twenty-five pigs, and a number of chickens that changes depending how many are eaten by bald eagles or raccoons. It's about as far from Silicon Valley as you can get. To make my regenerative farm work, I had to learn about soil, which is what puts minerals into plants. In turn, those plants put minerals into the animals that eat them, but only if the animals can actually digest and absorb the plants.

It turns out that it's easy to grow a mineral-deficient vegetable. Just try one of the tasteless cucumbers from a conventional grocer.

By comparison, try one grown in mineral-rich, healthy soil. I can tell you, it's not even the same food, and your body knows. Your MeatOS sends a signal to you that the food tastes better because it senses the nutrients. I discovered that it is also easy to feed livestock incorrectly. Certain plants will make a sheep or a pig sick. If they don't eat the right foods and minerals, they will develop split hooves. Bad feeding results in eggs that won't hatch, infertile sheep, fragile pigs that die when it's cold. Farming is a master class in learning how food affects animals, including us.

You come to realize and respect that ruminants (cows, sheep, and similar animals) are amazingly good at taking foods we can't eat, processing them, and turning them into nutritious foods we can eat (such as rib eye steak). But if we attempt to go to the source and eat plants as a cow does, the results can be disastrous. That's because most plants conspire to steal your minerals or harm you in other ways. Cows have stomachs and a digestive system set up to disable plant compounds that would otherwise steal their minerals. Humans don't. When we eat plants, we actually starve ourselves of raw materials. We add to the friction.

There you have it. Your veggies didn't grow in carefully cultivated, mineral-balanced soil unless you grew them yourself and you're a soil nerd. Your meat animals didn't eat mineral-rich plants unless they were pastured and grass fed. That means they ate the cheapest food possible, which means they got extra mineral-robbing antinutrients. Either way, you are in a bad situation because life doesn't work well without minerals. It will do its best to get by, but the whole point of this book is to help you do more than just get by. There's no sense in doing hard things such as exercise if you don't have the critical ingredients you need to benefit from them.

The first step toward getting the right resources is banishing antinutrients from your diet—and that takes work. Mineral deficiencies creep up on you, and it can take a while to see the results when you stop eating antinutrient-rich foods. The good news is, once you eliminate the foods that drain your resources, everything else becomes easier.

ENERGY ENEMY NUMBER ONE: PHYTIC ACID

The biggest cause of resource friction in your diet is a natural plant chemical called *phytic acid*.[1] Phytic acid is a potent antinutrient that binds to calcium, iron, magnesium, chromium, manganese, and zinc; when it binds to a mineral, it creates a compound known as *phytate*. Once that binding happens, your body can no longer process the mineral and you get stuck in a less functional, less energetic state. Phytic acid has become a much bigger problem in the last decade as Big Food companies have succeeded in selling "plant-based" foods as being good for you and the environment, even though the real reason is that they are high-profit, low-cost foods. Phytic acid was one of the five antinutrients I wrote about in my first big diet book, but it's an even bigger problem than was understood at the time.

The list of foods that are allegedly good for the environment and for you reads like a laundry list of high–phytic acid foods: nuts, seeds, beans, legumes, whole grains, soy, and corn. Humans don't make meaningful amounts of the enzyme that cancels out phytic acid, so we can't safely eat those foods unless we undertake arduous preparation techniques. Even worse, Big Food doesn't have any incentive to use complex processing techniques to reduce the phytic acid in the foods they sell. We're facing an epidemic of metabolic dysfunction and osteoporosis, and a huge amount of that can be traced to the increased phytic acid in "plant-based" diets in combination with our getting fewer minerals because of nutrient-depleted soil.

It's a perfect storm, nutritionally speaking. We are eating fewer minerals and at the same time eating more of the stuff that sucks them out of our bones. The lack of resources is dragging us down.

To operate efficiently and generate full energy, your cells require a full dose of *macro*, or "big," *minerals* (calcium, phosphorus, magnesium, sodium, potassium, chloride, and sulfur), along with *trace minerals* (iron, manganese, copper, iodine, zinc, cobalt, and selenium) and *ultratrace minerals*, which are biologically useful but found in only tiny quantities in Earth's crust. These minerals help build critical proteins, move chemicals into and out of cells, and transmit signals. Most of us are deficient in at least some of them, and phytic acid is one big reason why.

Reducing the amount of phytic acid in your diet is therefore crucial for having more energy. Just look at the problems caused by mineral deficiencies: Low levels of chromium, vanadium, and selenium are associated with diabetes. A shortage of molybdenum slows down one of the body's major detox systems. In general, you need minerals to make enzymes, the specialized proteins that facilitate the chemical reactions required to build cells and run your metabolism. Enzymes enable reactions that would normally require a lot of heat and energy to occur with very little energy. Take away the minerals, and your enzymes don't work properly, causing more friction. Suddenly, you are losing the components that enable the magic of life to happen.

Cutting down on phytic acid is especially important for women in their childbearing years. Guess what happens to a nutrient-depleted mother who doesn't have enough minerals in her body when she gets pregnant? Her body harvests minerals out of her skeleton, her liver, and other organs and redirect them to keeping her baby as healthy as possible. In fact, research has shown that one cause of postpartum depression is copper deficiency,[2] which is intimately linked to iron deficiency. There's a reason why animals commonly eat the placenta after birth; it's a way to get the minerals it contains back into their bodies. It's vitally important for new mothers to maintain their supplies of big minerals and trace minerals.

Phytic acid is insidious, and it is everywhere—in beans, seeds, cereal grains, and nuts. It almost seems as though Mother Nature is conspiring against you to prevent you from eating plants. Actually, that pretty much *is* what's going on. Plants don't want you to eat their babies, so seeds in all forms tend to be high in phytic acid. If you eat too many plant babies, you will be weakened and produce fewer offspring. Your species population will decline, which is good for the plants but bad for you. The only kind of seeds that evolved to be eaten are the digestive-resistant ones that are embedded in fruit, such as cherries. Animals consume the sweet fruit and poop or spit out the seed, which then grows in a new location. That process is beneficial to the plants, so fruits are generally much lower in toxins than seeds are, but only when the fruit is ripe and the seed is ready to be planted.

Although scientists have only recently begun to understand the chemical workings of phytic acid, our ancestors knew since prehistoric times that plants contain harmful compounds. The best way to eat grains and seeds (if you must) is to use traditional processing techniques, the kinds that few people do any more: sprout the seeds, then treat them with acid or ferment them, often for several days. Go to Central America, and look at how the locals used to prepare quinoa. Quinoa can wreck your gut—and it does when you eat it the way it's served now in the West—but in the local tradition, people fermented it for a couple days or pressure-cooked it to make it easier for digestion. They didn't know the exact chemical reason why those preparations helped, but what they were doing was breaking down the phytic acid in the plants.

Keeping with the traditional approach, you want to stay away from wheat. If you're going to eat a grain, a much better choice is rye, the grain that has the highest amount of phytase, the enzyme that breaks down phytate. You can use it to make sourdough bread, because the sourdough fermentation process breaks down phytic acid the most. If you follow all the traditional preparations and end up eating a proper rye sourdough, you'll probably handle it far better than you do white bread. It's not going to suck all the minerals out of your body the way that even wheat sourdough does. That's one way to go. The other, much simpler way is to cut way back on the amount of bread and other grain-based foods in your diet.

You know the falafel and hummus served in all those health food restaurants? They're mineral and energy sucks. Legumes, especially chickpeas, are superhigh in phytic acid. Peas range from 0.2 to 1.2 grams of phytic acid in every 100 grams. Dried peas are even worse because the drying concentrates the bad stuff; they contain between 0.3 and 3 grams of phytic acid per 100 gram serving. Be especially wary of peanuts, which not only punch you in the face with phytic acid but also contain unhealthy very-long-chain fatty acids (VLCFAs) and *lectins*,[3] another notable antinutrient. Like phytic acid, certain types of lectins interfere with the absorption of minerals, including phosphorus, iron, calcium, and zinc.

Nuts are a staple of alleged health food diets, including paleo, but they, too, can be a major source of phytic acid—up to 9 grams per 100 gram serving. The worst of the common nuts that people eat is the almond; the one that's likely the lowest is the pine nut. Macadamias are low, too. Walnuts can range from 0.2 gram up to almost 7 grams. Most normal people don't eat enormous servings of nuts at one sitting, though, unless you're like me when I was a vegan. I ate a ton of phytic-acid rich nuts and seeds. As they leached away my mineral reserves, I developed temperature-sensitive teeth and eventually cracked two teeth. This is common even in well-formulated vegan diets containing lots of minerals. The nuts, seeds, soy, and other antinutrients cancel out and inhibit the absorption of any minerals that might be present in the plants—if they were even grown on soil that contained minerals in the first place.

Speaking of soy, soybeans contain a significant 1 to 1.5 grams of phytic acid per serving, and people eat a lot of them. If you thought you were being healthy when you ate those walnut-soy burgers, think again. Actually, you were chowing down on antinutrients from two different sources.

A related mineral robber in plants is a defense chemical called *oxalic acid*,[4] found in high levels in raw spinach, kale, rhubarb, and many other common plant foods. Oxalic acid is an increasing cause of kidney stones. When you eat spinach, kale, and other "leafy greens," the vegan community tells you that you are getting lots of minerals. It doesn't tell you that the oxalic acid in those greens is stealing the minerals back or that the iron that might be in the spinach is approximately 2 percent absorbable compared to the iron you get from red meat. When you eat oxalic acid, it binds to calcium in your body to form tiny, razor-sharp crystals that circulate in your blood and get into your joints, your kidneys, and other tissues. You don't get the benefit of the calcium. Instead, you get a heightened risk of inflammation.

TOP ANTINUTRIENT FOODS

Grains (wheat, barley, sorghum, oats, corn)

Legumes (soy, chickpeas, beans, lentils, peas, peanuts)

Nuts and seeds (almonds, walnuts, sesame seeds, Brazil nuts)

Uncooked leafy greens (raw kale, spinach, chard)

Nightshade vegetables (white potatoes, eggplant, peppers, tomatoes)

LECTINS: THE INFLAMMATORY ENERGY DRAIN

Like phytic acid, lectins are defense compounds that plants make because they don't want to be eaten. They are a family of proteins that stick to certain types of carbohydrates in the body. Lectins are biologically common, and most forms are harmless or even useful. The trouble comes from specific types of lectins in common plant-based foods. If you have the wrong kind of sugar in your tissues and you eat the wrong kind of plant containing the wrong types of lectins, the plant compounds will bind to the natural sugars in your body and create chronic inflammation.

My friend Steven Gundry, a cardiac surgeon turned medical author, wrote a book called *The Plant Paradox: The Hidden Dangers in "Healthy" Foods That Cause Disease and Weight Gain* that focused on the complex role of lectins. He explains how lectins cause a condition known as leaky gut syndrome by creating microscopic holes in your intestinal wall. Once your gut wall is permeable, inflammatory compounds can leak out of your GI tract and into your blood. Now your MeatOS is going to partition off some amount of energy in order to deal with the inflammation and fix the damage, and it's going to increase the amount of biological anxiety you feel. In other words, you will have more stress and less energy. For some people, like me, certain lectins cause a massive stress response and an immediate drop in energy. More commonly, it's a delayed response that chips away at your energy over time.

Lectins can be especially troublesome because of their effect on glucosamine, a carbohydrate found in your joints. Nightshade vegetables make a lectin that sticks to glucosamine. The name *nightshade* probably makes you think of deadly nightshade, the most famous member of the family, but white potatoes, eggplants, peppers, and tomatoes are also nightshade vegetables. Because of that connection, they were considered deadly even to touch when they were first imported to Europe from the Americas. Nightshade vegetables were ornamental—until some hungry person ate a potato, probably.

Today, nightshade vegetables are dietary staples, which is a problem if you (like me) are susceptible to nightshade lectins, which can lead to joint pain and inflammation.[5] I grew up in New Mexico eating New Mexico green chili. It's one of my favorite foods. I'd slice habaneros and put them on everything. I also grew up with arthritis and pain throughout my body, without realizing the connection. When I quit eating the nightshade family, my pain went away. If I eat nightshades today, the pain comes back. Depending on your genetics, you may be able to eat eggplants but not bell peppers or some other combination. The easiest way to find out if you are sensitive is simply not to eat anything containing lectins for a little while and notice how you feel.

Grains are another source of antinutrient friction. Along with their phytic acid and lectins, they are a disaster. You know why so many diet guides tell you to eat grains? Because they contain fiber. But when you eat grains—especially whole grains, which contain intact phytic acid and lectin—you're getting a maximum dose of the bad stuff. You are blocking your body's ability to access any minerals that might be locked away in the fibrous outer layers of the whole wheat or the brown rice. The whole thing is a scam. Just say no.

HIGH-LECTIN FOODS

Nightshade vegetables (white potatoes, eggplant, peppers, tomatoes)

Grains (wheat, barley, sorghum, oats, corn)

Legumes (soy, chickpeas, beans, lentils, peas, peanuts)

| OMEGA-6: THE BAD FATTY ACID |

The name sounds similar to that of beneficial omega-3 fatty acids, and the two are chemically related, but omega-6 fatty acids are a drag on both your resources and your health. We know that eating more than a small dose can cause inflammation, and the stuff is everywhere. The modern Western diet is heavily overloaded with omega-6 fatty acids.[6] The oils containing them are unstable, so they readily create free radicals, especially when exposed to heat. Cooking, microwaving, or frying them will speed up and increase destructive oxidation processes. The free radicals created by oxidized omega-6 fats damage the DNA in your cells, inflame your heart tissue,[7] and may raise your risk of developing several types of cancer, including breast cancer.[8]

The fat stores in your brain are exquisitely sensitive to the amount of omega-6 fat that you eat. A brain that is rich in omega-3 fats will perform better than one that is rich in omega-6 fats. The standard reason you will hear for this is "inflammation." The reality is more complex and nefarious. Inside the mitochondrial membrane is an important compound called *cardiolipin*, which helps your mitochondria move electrons around in the body. The more omega-6 fat you eat, the more your body includes in your cardiolipin and the less effective the electron transport chain becomes.[9] Babies have the highest biological energy levels, and their cardiolipin is 100 percent saturated, with no omega-6 to speak of. As a final insult, omega-6 interferes with brain metabolism. It is the kind of fat that puts you to sleep—literally, it is the kind used by hibernating animals. When you eat a lot of omega-6 fats, you are in hibernation mode, not action mode.

The most common form of omega-6 is linoleic acid, found in high amounts in processed oils such as corn oil, soybean oil, safflower oil, cottonseed oil, and sunflower oil. It is also abundant in certain whole foods such as poultry and some nuts and seeds. Vegetable and seed oils are cheap to produce, so many companies use them in processed foods from crackers to dairy creamer to frozen pizza. Soybean oil is so overused that it accounts for about 20 percent of the calories in the typical American diet. Many of these common oils are also derived from genetically modified plants and are extracted with toxic solvents.

I pointed out the problems with omega-6 fatty acids in my earlier book *The Bulletproof Diet: Lose Up to a Pound a Day, Reclaim Energy and Focus, Upgrade Your Life*, but since then the scientific evidence has grown more ominous.[10] It now seems that eating food fried in canola oil causes more inflammation, and for longer, than smoking a cigarette does. If you care about your health and want to perform at your best, ditch them both. In the Sydney Diet Heart Study, a seven-year randomized controlled trial in humans, increased consumption of vegetable oil was found to increase the risk of premature death by 62 percent—more than physical inactivity, heavy drinking, moderate smoking, increased sugar, processed meat, or excessive sodium.[11]

Although we call them all "fats," animal fats do very different things in your body than omega-6 fatty acids do. Most animal fats, such as beef tallow and butter, contain mostly saturated fats, and because of this, they are difficult to oxidize. When an oxygen atom sticks to a molecule, it creates a free electron that bounces around looking for something else to stick to. Those reactive molecules attack the cells in your body, creating reactive oxidative stress and/or inflammation. The body can handle a certain amount of inflammation, but after that it loses its ability to protect itself and you start getting oxidative damage. Omega-6 fats have many spots where oxygen could stick; in technical terms, they are *polyunsaturated*, with multiple places where oxygen can stick. That makes them a potential source of inflammation that your MeatOS is going to have to deal with—by stealing your energy.

To be clear, small quantities of naturally occurring omega-6 fats are fine and even healthy. Your body actually needs a small amount of omega-6 fats to create a response to injuries or pathogens. Many healthy foods, such as beef and avocados, contain small amounts of omega-6 fats, but they usually come in conjunction with other beneficial compounds.[12] For example, olive oil contains omega-6 fats that are packaged with oleic acid, a type of omega-9 fat. Studies show that increasing the amount of oleic acid in your cardiolipin is associated with better health outcomes and better mitochondrial function. Olive oil also contains many antioxidants that likely work with the omega-6 fatty acids to protect them from oxidative damage. Moderate consumption of omega-6 fats, such as a tablespoon of olive oil or a few slices of avocado per day, is fine. What you really want to focus on is replacing refined high-omega-6 plant fats and oils with healthier animal fats.

Soon you won't even have to go to animal fats. One of my most exciting angel investment companies is called Zero Acre Farms. Its founders have discovered a way to use almost any plant material to create a wide variety of fats. They're developing formulas that can replace soy, corn, and canola oil with lower-cost, far healthier oils. Fortunately, the world is starting to catch up with the science about fatty acids. A few years ago, the US Food and Drug Administration partially banned another harmful type of dietary fat, trans fat, in processed foods.[13] Omega-6s are still everywhere, but you can learn to cut back on them significantly. When you remove that friction, your MeatOS will start to run more quickly and be easier to program.

HIGH-OMEGA-6 FOODS

Processed vegetable and seed oils (corn oil, sunflower oil, soybean oil, cottonseed oil, canola oil)

Nuts (walnuts, pecans, Brazil nuts)

Seeds (flax, sunflower, sesame, pumpkin)

Poultry fat (chicken, turkey, duck)

Conventionally raised pork fat/lard

SNEAKY SOURCES OF OMEGA - 6 FATS

Dressings, sauces, and condiments

Processed and packaged snacks

Omega-6 and Omega-3 Content in Processed Vegetable and Seed Oils

Ingredient	Omega-6 Content	Omega-3 Content
Safflower	75%	0%
Sunflower	65%	0%
Corn	54%	0%
Cottonseed	50%	0%
Sesame	42%	0%
Peanut	32%	0%
Soybean	51%	7%
Canola	20%	9%
Walnut	52%	10%
Flaxseed	14%	57%
Fish (wild)	0%	100%

USDA National Nutrient Database, USDA, fdc.nal.usda.gov

| HISTAMINES: THE HANGOVER MOLECULES |

Everyone has heard of histamine because antihistamines are what you take when you have seasonal allergies. Almost no one knows that foods, particularly fermented foods and leftovers, can contain enough histamine to trigger other types of allergies. Histamine-rich foods such as fish, pork, and fermented soy products can throw a huge wrench into your program to upgrade your biology.

I've always been aware of histamines, because allergies have plagued me at various points in my life. As a kid, I went through countless futile exercises of injecting various allergens into a grid on my back to see what would cause my skin to exhibit a histamine reaction. Doctors couldn't figure out why I was always coughing and having other weird symptoms. They looked at tree pollen and weeds, but back then they never tested for mold, which is a major cause of allergies. After shrugging their collective shoulders, they put me onto allergy medication and sent me on my way, even though the medication didn't work. Over time, I figured out how to biohack my allergic response so I don't get the brain fog my allergies used to cause. That is, until contracting COVID caused a mild allergic cough to get—and stay—much worse. Two years into the pandemic, many people are reporting the worst allergy season they can remember. The fact that everyone wore masks and cleaned excessively early on surely contributed to some of the increase; we know that regular exposure to mild allergens keeps your immune system strong. But what I saw from my own worsening of symptoms matches what I am hearing from people who got COVID and recovered or who were vaccinated: allergies are off the charts. Knowing that histamine causes allergic symptoms and that it's one of the most common food toxins, I did some deeper research into the ways histamine interferes with the MeatOS.

It turns out that making you sleepy is far from the worst thing histamine does. It can also set off your mast cells, immune cells that are like land mines embedded throughout the body. But since they are close together, imagine what happens when one goes off: it sets

off the others around it, which creates a chain reaction throughout the body. Triggered by histamine, your mast cells not only release more histamine but also about a hundred other inflammatory
chemicals that slow you down and make you inflamed and itchy.
Overly sensitive mast cells cause a huge range of problems in the
body: low energy, hot flashes, hives, brain fog, sleepiness, joint
pain, back pain, rashes, eczema, anxiety, and psychological problems. Over the next ten years, as a result of COVID, I expect that
we will come to understand a lot more about mast cell activation.
I suspect that it plays a role in a huge number of neurological and
digestive issues.

If you are fortunate, you naturally have a large population of
histamine-degrading bacteria in your gut and enough histamine-
degrading enzyme (an enzyme called diamine oxidase, or DAO)
in your liver that you can eat histamine-rich foods and feel nothing. Most people do feel it, however. One common reaction is feeling tired, as the excess histamine goes to work in your brain. You
might eat a food and afterward feel like, Oh my God, is that a
fish hangover? Many people respond to histamines in their food
by puffing up, getting a temporary muffin top. They also feel tired
and cranky, suffer from headaches, become light sensitive, develop
a runny nose, and catch colds—all because they happen to have
sensitive mast cells. If you are histamine sensitive and you are about
to eat a food such as cured pork, which is likely to be high in histamine, you can take a hit of commercially available DAO enzyme,
which will help your body break down the histamine. Or you could
do what I do when I'm out of DAO enzyme: I take a quarter of a
Benadryl if I'm going to eat leftovers, and I feel just fine. Claritin
also works.

It is hard to avoid all histamine, but you can test your sensitivity
to figure out how careful you need to be. Just have a Benadryl on
hand in case you're going to eat something containing fish sauce,
soy sauce, or cured pork or fish. If you feel great afterward, you
probably have a high enough DAO level and histamine won't interfere with your body's operations. If you cough or get heartburn,
your nose runs, your eyes water, or you feel nauseous, you are prob-

ably sensitive. If you crave sugar and feel that you must nap, you're sensitive. If your skin suddenly feels itchy, you're sensitive. Or if you feel that the sun or indoor lights are too bright. If you wake up the next morning with stiff joints, a sore back and hips, or hives, feeling groggy, you are sensitive. Lowering the amount of histamine in your diet will magically increase your energy and lower your cravings.

HIGH-HISTAMINE FOODS TO AVOID

Alcohol

Pickled and canned foods, such as sauerkraut

Matured cheeses

Smoked meat: salami, ham, sausages

Shellfish, especially scallops

Cured fish and fish sauce

Beans and other pulses: chickpeas, soybeans, peanuts

Chocolate (varies widely)

Leftover foods

GLYPHOSATE: THE OUT-OF-CONTROL WEED KILLER

Unlike phytates, lectins, and histamines, glyphosate is an obstacle that we have created for ourselves. It's an artificial herbicide (weed killer) that is ubiquitous in modern farming, especially wheat grown in the United States. I've been warning about its dangers for years, even as the agricultural industry kept insisting that glyphosate was completely harmless to humans. In reality, it is a toxin that accumulates in fat and neurological tissues. Now we know that glyphosate disrupts your gut microbiome,[14] the healthy bacteria that live in your intestines, and has been tied to certain types of cancer.[15] Nevertheless, it is still used everywhere.

More concerningly, glyphosate has been shown to modify your mitochondrial membrane and decrease your energy production.[16] Anything that lowers your energy level also reduces your willpower. Specifically, glyphosate can switch your brain cells from aerobic (powerful) metabolism to anaerobic (less powerful).[17] In other studies, glyphosate clogged up the respiratory cycle of mitochondria in the livers of rats[18] and harmed energy production in human kidney cells.[19] And we now know that glyphosate is another antinutrient that binds to the essential minerals: iron, copper, zinc, manganese, calcium, and magnesium.[20]

In other words, glyphosate sabotages your MeatOS at multiple levels. It decreases your energy production, which increases your physical sense of anxiety and stress and leaves you with fewer resources to work with.

In addition to using glyphosate to kill weeds, farmers spray it on crops near the end of the growing season because it causes the wheat to ripen early, at a time that they can schedule. Glyphosate kills the plant, so the plant puts its last burst of energy into saving its babies—that is, it rushes to make all the wheat grains ripen before it dies. The problem is that the glyphosate then continues on into the foods that we eat. It also destroys topsoil by killing the bacteria that normally cycle carbon and renew the soil's organic chemistry. This has to stop. The companies that make glyphosate and the farmers who use it should be held accountable for what they're doing to our planet.

You can do your part by steering clear of foods that were made with glyphosate-treated crops. Unfortunately, glyphosate isn't easy to avoid. Choosing organic food is a good first step, but farmers spray so much that many organic foods, even organic California wine, contain concerning amounts of glyphosate. Since grains usually contain the highest amounts of glyphosate residues,[21] cutting down the amount of them that you eat is the most effective way to start. If a grain isn't organic, don't eat it.

**HOW TO AVOID GLYPHOSATE
IN YOUR FOODS**

Shop at a farmers market, and know your farmer.

Purchase organic produce.

Eat certified organic grass-fed and -finished beef.

Wash all vegetables, even organic ones, very well.

Limit your consumption of grains; whatever grains you do eat should be organic only.

CHICKEN, FAKE MEAT, AND LOW-QUALITY PROTEIN

I'm done with the problematic plants, but unfortunately I'm not nearly done with the list of foods that hold you back. Another friction-causing food is a staple of the modern American diet: the industrial chicken.

On my farm, the chickens wander around outside and eat bugs and worms, the diet they evolved to eat; they take nine months to reach full size. They are nothing at all like the monstrosities that show up in fast-food restaurants or packaged in little Styrofoam trays. Those birds are raised on factory farms, never see sunlight, can barely move, are crowded together, and are fed a fattening diet of corn and soy. All of that terrible treatment is designed to maximize profit: a factory chicken takes just six *weeks* to reach full size.

By their nature, chickens make a lot of omega-6 and other inflammatory fats that sap your energy and degrade your health when you consume their meat. Eating omega-6 fats from chicken lowers your cellular power levels, which creates anxiety in your MeatOS. Animals that are crowded together, chickens in particular, also make a lot of amyloid protein—abnormal proteins that tend to clump together in the body. Amyloid protein clogs up the lysosomes,[22] the

garbage incinerators in living cells that destroy the buildup of toxins. If those incinerators get clogged up, your cells become less efficient. Other poultry, such as turkey and duck, are also high in
omega-6 fatty acids, so I recommend minimizing your consumption
of these as well.

Pigs are better for you, but I still recommend that you limit your
pork consumption. The good things about pigs are they can eat almost anything and that pig fat is nutritionally very similar to human
fat. If you feed a pig a healthy diet, it will produce fat that is higher
in healthy saturated fat and lower in omega-6 fat. On the other
hand, if you eat the meat of a grain-fed industrial pig, it is still going to contain lots of omega-6 fat, which is bad for you in excess. I
raise my own pigs, which eat the right diet: mostly vegetables with
some milk and table scraps, including lots of lamb fat, thrown in.
This way, the pigs are high in saturated fat and good for you. I don't
expect that many of you are going to start your own pig farm, however. The next step down is to buy lean pork that didn't come from
an industrial farm. Your most practical option, though, is to keep
your consumption of pork modest.

Probably the most important protein to avoid is any type of plant-
based meat substitute. These highly processed items are full of ingredients that contribute to inflammation and aging, including genetically
modified soy, omega-6 oils, synthetic vitamins, refined sugars, and
flavors. Since they are mainly legume or grain based, they contain
tons of mineral-sucking antinutrients, too. They shouldn't even be
called meat substitutes, because plant proteins are much less bioavailable than animal proteins are, meaning that your body can't
use them nearly as well for building muscles and repairing tissues.[23]

You may be tempted to guzzle down shakes made with protein
powder, but most likely you're not doing yourself any good. Most
commercial whey protein powder is made from industrial dairy
products and is loaded with artificial sweeteners that wreck your
gut microbiome. Plant-based protein powders such as those made of
soy, pea, and rice protein are filled with antinutrient phytates and
lectins and tend to carry frightening amounts of heavy metals such
as lead, cadmium, and arsenic.[24]

If you're going to consume protein powder, your best bet is to go for a lab-tested, grass-fed and -finished raw whey protein concentrate or an organic, lab-tested defatted hemp protein.

Another important point about protein (animal or plant) is that you don't want to get too much of it relative to the amount of animal fat you consume. A high-protein, low-fat diet is quite harmful, because your body is designed to burn fats and carbs, not protein, as its main fuel sources. Your body can use protein to make energy if you aren't getting enough fat or carbs, but doing so involves multistep processes that are metabolically expensive, meaning that they suck up your energy, leaving you with less energy to kick ass in other areas of your life. Plus, when your body uses protein as a main source of energy, it creates harmful ammonia as a waste product.

Ideally you want to be burning carbs and fat most of the time and have backup protein that you use mostly as building blocks. For your personal biohack, you want to eliminate bad proteins but keep the good ones—and you want to make sure that you are getting plenty of good fats after you are done eliminating the bad ones.

LOW-QUALITY PROTEINS TO MINIMIZE
Chicken
Turkey
Duck
Conventionally raised pork

LOW-QUALITY PROTEINS TO AVOID
Plant-based "meat"
Plant-based protein powders
Most whey protein powders

| ARE YOU DRINKING THE WRONG KIND OF MILK? |

Another major obstacle in your diet comes from the modern dairy industry. When we allowed giant agribusinesses to turn dairy cows into a huge, mechanized industry, we did nasty things to our milk. We stopped letting cows eat grass, their natural diet, and started feeding them grain, along with antibiotics and sometimes hormones. The goal was to make the cows more efficient as milk-producing machines, and it worked.

A modern industrial cow can produce up to thirty gallons of milk a day. It converts grain effectively into milk; that milk just happens to contain a disruptive protein called A1 casein. There are two types of casein protein, A1 and A2. When you drink milk containing A1 casein, your digestive enzymes break it down into a by-product that can influence the nervous, endocrine, and immune systems by activating opioid receptors in your body.[25] Studies show that A1 dairy products also significantly increase inflammatory markers in the colon, which can lead to digestive symptoms.[26] In fact, many people who think they are lactose intolerant aren't actually reacting to lactose but instead to the casein in the milk.[27]

Whether or not your milk is A1 or A2 depends on the breed of cow from which it was sourced. Modern industrial cows produce A1 casein. To get the healthier A2 protein, you need to consume dairy products from Guernsey or Jersey cows or from sheep, goats, camels, or buffaloes. Choose your dairy products wisely, and know where your milk comes from.

A second problem with modern milk is pasteurization. Heating milk to a high temperature alters a lot of the bioactive proteins and peptides.[28] It cooks them so that they fold and become more inflammatory in the immune system.[29] These altered, or *denatured*, proteins also lose some of their natural immune-signaling benefits.[30] There are many studies showing that pasteurized milk is harmful, unlike raw milk, which still contains intact proteins and immune factors. There's been a huge regulatory war over raw milk because it can contain bacteria if it's not handled correctly. But raw dairy products have been shown to be phenomenally safe when handled appropriately.

Finally, there's homogenization. Back in the days when a milkman delivered fresh milk to your house, it arrived with a layer of cream on top because the cream and the milk naturally separated. Homogenization keeps the milk even and consistent by applying very high pressure and forcing raw milk through a microscopically small screen that breaks up the fat droplets. The natural-sized droplets of dairy fat have been shown in a lot of studies to be beneficial.[31] Homogenization turns them into something different.

Industrial dairy fat is broken into droplets much smaller than the ones that occur in nature. These micro–fat globules can go anywhere in the body, and they can even cross cell membranes. They no longer behave like food.[32] At the same time, the homogenization can damage the proteins in the milk,[33] on top of the damage done by the pasteurization. Unhomogenized milk always has a layer of cream on the top that you have to stir back in. When you consume it, you take in different sizes of fat that your body can process as building blocks or fuel, as opposed to microdroplets of fat, which carry proteins across the gut barrier.

The simplest way to take dairy friction out of your diet is to stop drinking industrial milk and instead support a small-scale dairy farm that sells raw milk and dairy products. There are also some more exotic options. Camel's milk is highly compatible with human biology, but it's almost impossible to find in the United States. (Full disclosure: I was an adviser to the one company importing raw camel's milk.) Sheep's milk is a good alternative. Sheep eat mostly grass, and sheep's milk also contains proteins that are compatible with humans'. Sheep's milk also has a better saturated fat ratio and contains more protein than cow's milk. In the years since I published *The Bulletproof Diet*, I have incorporated substantial amounts of sheep's milk cheese and yogurt into my diet. After sheep's milk, I would go to goat's milk.

Let's get real. You're probably not going to track down those other kinds of milk, but you definitely can and should get the industrial, homogenized, pasteurized milk out of your diet.

Look for organic, raw, grass-fed A2 milk, the best cow milk. After that I would choose organic, grass-fed A1 milk that has not been pasteurized and homogenized. Avoid A1 pasteurized homogenized milk, which is what you find in vitamin D–fortified supermarket milk.

TIPS ON CHOOSING DAIRY PRODUCTS

- Avoid A1 dairy products, and pasteurized and homogenized milk.
- Switch to grass-fed, organic raw A2 cow dairy products.
- Seek out dairy products from sheep, goats, buffaloes, or camels.

BIOGENIC TOXINS: THE MOLDS OF MISERY

This is an area where I wish I'd known what to avoid many years ago. I've mentioned my long and miserable history with mold toxins, aka mycotoxins. There are about two hundred types of them, and zearalenone is one of the worst.[34] It is such a powerful obesity trigger, or obesogen, that it is concentrated, packaged, and sold for fattening purposes. In commercial form it's called zeranol and formed into a waxy pellet that farmers can put into a cow's ear.

The toxin from zearalenone slowly drips into the cow's ear and into its bloodstream, where it acts like a much more powerful version of estrogen. That cow will then get fat on 30 percent less calories. If zearalenone gets into its milk, it will do the same thing to humans.[35] It's bad enough being accidentally exposed to a building that contains mold toxins. Don't voluntarily expose yourself to diabetic cows that have been hormone treated that way. It's one more reason to avoid both factory-farm meat and factory-farm dairy products.

Aflatoxins and ochratoxin A are two other types of mycotoxins that wreak havoc on your health.[36] Ochratoxin A can cause kidney damage, and aflatoxins can cause DNA damage and liver cancer in humans. Common sources are grains, corn, peanuts, nuts, non-lab-tested coffee, low-quality chocolate, and alcoholic beverages. These items are especially susceptible to mycotoxin contamination because of the way they are processed and stored. It is best to avoid these foods if you're trying to upgrade your MeatOS.

HIGHEST SOURCES OF DIETARY MYCOTOXINS

Conventional dairy products

Grains

Nuts

Non-lab-tested coffee

Low-quality chocolate

Alcoholic beverages

FISH: NOT SUCH A SMART FOOD AFTER ALL

Many people will be surprised to see fish on my list of foods to avoid, but fish are moving down in the ranking of healthy foods. One reason why is that a lot of people are going on pescatarian diets that avoid meat but include fish, thinking that the fish will make up for the nutritional value of the meat. Wrong. Fish don't contain the right minerals or the right fat. Yes, you need some omega-3 fat in your diet, but you don't need that much of it.

A second problem is that the amount of microplastics in fish has gone through the roof over the past fifteen years;[37] microplastics can carry toxins into the body, and they themselves may leach toxic compounds.[38] Likewise, the amount of mercury in fish has gone through the roof. Humans have done a fantastic job of trashing our oceans with health-threatening materials and compounds, including the persistent industrial chemicals known as polychlorinated biphenyls, or PCBs. These days, when you are in a restaurant, you have to consider carefully: If it is serving industrial beef, industrial chicken, or industrial fish, which is the worst? Which one should you avoid first?

Personally, I would pass on all three and order some veggies and rice, thank you very much. Make it white rice, and put some grass-fed butter on it. If the restaurant doesn't have good butter, I bring my own. (Yes, I'm the kind of guy who travels with his own butter.) I avoid fish of unknown origin. I do eat some fish, but mostly sushi-grade fish from restaurants that are careful with what they serve. Even then, I take supplements to bind the toxic metals when I eat fish, because heavy-metal contamination is a global ocean problem. I also take supplements to kill the parasites that are commonly found in raw fish.

When I have a choice, I eat bivalves. Oysters, clams, and mussels are high in nutrients, including copper, zinc, and other important minerals. If you are going to eat fish, go for younger fish that have had less time to accumulate plastics and toxins. Salmon, especially sockeye salmon, is generally better than other fish, because they live only two years and spend most of that time in fresh water. Do not eat farmed fish, especially from the coast of British Columbia. For years, I've been railing against the fact that Atlantic salmon farmed with antibiotics are killing local wild fish stocks. Now it's been proven beyond a shadow of a doubt, and the Environmental Working Group has announced that farmed fish from the west coast of Canada are no longer considered environmentally sustainable.

SEAFOOD TO AVOID

Tilapia

Farmed salmon

Shrimp

Shark

Tilefish

Swordfish

Tuna

| ARTIFICIAL ADDITIVES: WHAT A HEADACHE |

You probably already have an intuitive feeling that you want to avoid artificial sweeteners, artificial flavors, and artificial colors, but you may not know exactly why. One big reason is that they are highly disruptive of your gut bacteria. Researchers have run gut microbiome tests that have documented their ill effects.[39] Sucralose is particularly hard on your gut bacteria. Other additives, especially artificial colors and some artificial sweeteners, affect the brain. Commercial red and blue food dyes have been linked to a number of brain problems, especially in kids. Aspartame (Nutra-Sweet), in particular, has been tied to food cravings, which explains why ostensibly nonfattening sweeteners actually end up increasing obesity and sapping our energy. Aspartame is also known to cause headaches and affect the firing of neurons in the brain.

There has been a lot of debate over monosodium glutamate (MSG) over the years. To me, the scientific findings are clear. In the brain, an excitatory neurotransmitter called glutamate triggers neurons to fire. Too much glutamate can overexcite neurons, even to the point of death.[40] The glutamate in MSG hits that same trigger when it crosses the blood-brain barrier, which is why so many people get headaches or need to drink a lot of water after consuming MSG. It is widely used in fast foods, snack foods, processed meats, canned soups, condiments, and seasoning blends, so look carefully at ingredient labels, although it isn't always listed. Your better bet is to avoid these categories of foods entirely, especially since most of them are nutrient poor and full of compounds you don't want to be eating anyway.

BIGGEST ARTIFICIAL ADDITIVE OFFENDERS
Aspartame
Sucralose
Ace-K (acesulfame potassium)
Artificial flavors
Artificial colors
MSG (monosodium glutamate)
Yeast extract, autolyzed yeast extract

3

LOAD UP ON RAW MINERALS

Do you ever feel as though you're on a hero's journey? There are great things you want to do in your life: maybe starting a company, maybe creating art, maybe raising a family—or any combination of those things. You picture yourself taking on enormous challenges, powering through them, and succeeding triumphantly. If you've failed in the past, you psych yourself up to try again. If you've succeeded, there are more challenges ahead, waiting to be conquered. It's kind of a treadmill.

That all sounds noble enough, but I have a dirty secret: the hero's journey is a pain in the ass.

One of the most famous heroes in history is Pheidippides, the Greek messenger who ran twenty-six miles to let the rulers of Athens know that their forces had defeated the Persians in the Battle of Marathon. It was an amazing achievement, but there's one detail that people typically leave out of the story: Pheidippides collapsed and dropped dead at the end of his run. The real lesson of hero legends is that you don't want to be a hero. The heroes struggled because they didn't have a guide, and we celebrate the struggle because—well—our culture says we're supposed to. In reality, if you learn their lesson, they suffered so that you don't have to.

I want to save you from feeling that your life goals require painful, hard-won victories. You shouldn't have to experience the kind of misery that I went through before I understood the workings of the MeatOS and the ways to manipulate the body's laziness principle.

Avoiding friction is the essential first step. Next you want to load up on the raw materials that will ease you along your path. I'll be your guide, so that you don't have to come anywhere near the rock bottom that I did.

A big turning point for me—a moment when I realized I had stumbled onto a hero's journey and it was absolutely miserable—came when I was getting a business degree at the Wharton School. I was working a full-time job while going to the hardest business school in the country full-time. I was also dealing with a bad metabolism and eating a vegan diet that was depleting me. I looked around at the end of my first semester and thought, "I'm not going to make it." I'd sit down to take a test and focus as hard as I could, but I couldn't get my energy up. By the second question I was writing gobbledygook. I just felt slow. Out of desperation, I went to the world-renowned Amen Clinics, where the doctors scanned my brain and found that its metabolism was broken. There were big gaps in my brain with no electrical activity. It turned out that I had toxin-induced brain damage.

I was relieved by the diagnosis, because that meant that I wasn't as stupid and lazy as I had started to believe I was; it was just that my hardware was broken. I began to believe that maybe I could find ways to get my energy back. I was an extreme case, but every reader of this book can benefit from some version of this awakening. Do you feel emotionally blocked? Burned-out? Are you struggling with your weight? Not getting results in the gym? It's not because you're weak, it's because you have a hardware problem, and you can fix that. You don't have to suffer more to get better. Your number one task is to get your energy back. Then you will have your motivation and willpower back as well.

Once I realized what was broken in my body, I started studying the foods necessary to fix the problem. It took a long time, but eventually I restored my energy. Along the way, I discovered that amazing things happen when you eat the right foods and you have more energy. Not only do you get more done without that agonizing heroic feeling, but you also become kinder. I was talking to my parents, and they said, "Wow, you're a much nicer person. Your personality changed." That's a big payoff of eliminating the bad foods in your diet and loading up on the good ones. When you can barely get through the day, it's hard to have the energy to be patient with others, but when life doesn't feel like such a struggle, kindness comes much more easily.

I'm telling you as someone who is healthier now than I ever have been: if I could claw my way back, you can, too. I'll share my insights to save you time and effort. Choose the right fats and proteins. Get the right vitamins and minerals. Eat only the right plants. Give your body the right resources. You will have the electrical system of a young person, and you'll have the wisdom of an old person. You will be able to snap out of the baseline state of laziness whenever you need to spring into action. You'll feel better, you'll perform better, and you'll look better, too. No hero's journey will be needed.

WHAT YOU DO NEED

In the last chapter you learned about what you don't need—how to remove the friction that holds you back. Now let's dig into the basics of the things you do need to get into your diet so that you will have enough energy and building blocks to move forward. The guidelines here build on ideas that I have been developing for the past two decades, with significant updates based on the latest dietary research as well as my ongoing experiments in biohacking. Your to-do list starts with the right vitamins, minerals, and fats.

Fat-Soluble Vitamins

Everyone has heard of vitamins, but not many people know that there are only four kinds that dissolve in fat instead of in water—D, A, K, and E—and that collectively they are the most critical vitamins for the smooth operation of your MeatOS. And most people have no idea that if they are eating a standard Western diet, they are likely not getting sufficient amounts of "vitamin DAKE," meaning that their basal metabolism is broken. You need these vitamins to shuttle minerals around so that your body can build proteins, generate energy, and move electrical signals.

When you finally consume enough fat-soluble vitamins, magic happens. These vitamins don't dissolve in water, so they are best absorbed by the body if you eat them along with fatty foods. It's no coincidence that these vitamins are also most abundant in high-fat foods.

Vitamin D is first in my DAKE shorthand because it is the most important of the fat-soluble vitamins; it's one of the most important nutrients of all, in fact.[1] It helps you sleep well. It regulates your immune system and keeps inflammation under control. It promotes bone formation and assists key hormones, including estrogen and testosterone. Your body can synthesize its own vitamin D when your skin is exposed to sunlight, but unless you are sunning yourself in the nude all the time (no questions here), you need to get more of it.

Vitamin A facilitates many metabolic functions in the body. It also helps maintain your immune system and is essential for the proper operation of the light-sensing cells in your eyes. The beta-carotene in carrots can be converted to vitamin A, which is probably the source of the myth that eating carrots will make your vision sharper. In reality, your body struggles to transform beta-carotene into vitamin A. Liver, fish liver oil, and butter are better sources.

Vitamin K is two different vitamins, actually: K1 and K2. The critical one is K2, also known as menaquinone, which helps your body process calcium. If you don't consume enough K2, excess calcium can build up in your arteries, leading to a heightened risk of heart attacks and arteriosclerosis. K2 also helps to maintain bone density and stave off osteoporosis.

Vitamin E is an important antioxidant. It helps mop up highly reactive, oxygen-dangling free radical molecules in the body before they can attack and damage your cells. The oxidizing effects of free radicals contribute to the overall aging and degradation of the body. Vitamin E also functions as a blood thinner, reducing the risk of unwanted clotting. Unlike the other fat-soluble vitamins, E is commonly found in plant-based foods.

Minerals

In chapter 2, I introduced you to the three main categories of minerals required for proper enzyme function, so that your body can run its chemical machinery and generate energy efficiently.[2] Almost everyone is deficient in them, too.

Macro, or "big," minerals are the building blocks for your body. The key big minerals are calcium, magnesium, sodium, potassium, phosphorus, and sulfur. You have a lot of them in your body: about 1,000 grams (two pounds) of calcium in the average person. That means you need to replace them in high doses, larger than 100 mg per day.[3]

Although big minerals contribute to the structure of your body, they are also used in energy transfer. The calcium channel is an integral part of the electrical system of the body. Despite all the noise you hear about too much sodium in our diet, you probably need more sodium than you are getting. You might want to increase your salt intake; if you're suffering from brain fog, sometimes getting a little more salt in your diet changes everything. Stay away from refined iodized salt and choose sea salt, which is a great source of sodium and some trace minerals.

You want to choose foods that will provide more of the other minerals as well. Even then, you will almost certainly need to supplement; I'll tell you all about that in the next chapter.

Mesominerals really include just one element: iron. It's a double-edged sword. You need enough iron in your diet, because you need it to make hemoglobin, the oxygen-carrying molecule in blood. Without enough iron, you become anemic. Many women of childbearing age don't get enough iron, which can lead to complications if they get pregnant. On the other hand, too much iron in your blood will contribute to oxidative stress and premature aging.[4]

Trace minerals are present in vanishingly small quantities in the body, but they are nevertheless essential to the chemical reactions of life. The minerals in this category are: molybdenum, zinc, iodine, selenium, cobalt, copper, fluorine, vanadium, and manganese. Broadly speaking, the job of the trace minerals is to create enzymes, the molecules that facilitate vital biochemical reactions in your body. Trace minerals are analogous to the elements used for semiconductor doping in the electronics industry. Years ago, electrical engineers figured out how to improve the flow of electrons through silicon by adding tiny, tiny amounts of trace metals. That process, called doping, made it possible to build chips that are much smaller and operate much more efficiently. You wouldn't have your cool smartphone without it. Likewise, enzymes enable chemical processes to happen using far less energy than they would normally. Your metabolism could not run without them.

Ultratrace minerals are probably important for health, too. They have just not yet been quantified or their biological role precisely identified. That's why I've created a drink called Danger Coffee, which contains fifty different minerals from decomposed plants. There are likely to be unknown important nutrients in the mix.

Saturated Fats

Let's face it, you are made of meat. From a structural perspective, you want to eat the raw materials needed to make more meat so that your body can get rid of your old cells and create newer, healthier ones. In particular, you want to take in a lot of animal fat. There are three kinds of membranes in the body, and they're all made of fat. From an energy perspective, you want to eat foods that easily generate the electricity that runs your metabolism. The easiest foods to make electricity from are fat and soluble fiber, which gut bacteria will turn into a fat that the body can metabolize.

You therefore need a lot of high-quality fat in your diet. That's fundamental for having the right resources to do all of your other health hacks. At least 50 percent of the fat in your diet should be saturated fat such as stearic acid, one of the most common fats in tallow and lard, because that's what your body is made out of. Saturated fats, you may recall, are healthy also because they have no place for troublesome solitary oxygen molecules to latch on to. It's fine to include some monounsaturated fats in your diet as well, even though they can transport a little bit of reactive oxygen through the body. You just want to keep things in balance. A good target is 75 percent saturated fat and the rest monounsaturated fats, with just a bit of omega-3 fatty acids (associated with a reduced risk of cardiovascular disease). Remember that you are aiming to avoid excess amounts of omega-6 fatty acids and polyunsaturated fats.

Protein

The workhorse molecule of your body performs a wide array of vital functions. How important are they? The name comes from the Greek word *prōteios*, which means "holding first place." Proteins give structure to your tissues, organs, and muscles. They form antibodies to fight off infection. They make hemoglobin to carry oxygen around the body in the blood. They create hormones. Enzymes are a specialized class of protein. If you are not getting enough high-quality protein, you may experience swelling in your body and broken nails; you may get sick more often and notice that it takes you longer to heal; you may suffer from depression, weakness, and fatigue.

Proteins are constructed from smaller molecular building blocks called amino acids. There are at least twenty amino acids found in nature. Nine of those amino acids are essential to your diet, since your body can't manufacture them. The crucial nine are histidine, isoleucine, leucine, lysine, methionine, phenylalanine, threonine, tryptophan, and valine.

You can think of how protein is structured this way: The amino acids are like letters, put two or more of the letters together and you form a "word"; these are subprotein units known as *peptides*. Then string the words (peptides) together in a "sentence," and you get a protein. That protein can say something scary such as "Die now" if you just consumed shellfish toxin. On the other hand, it can say helpful things such as "Grow muscles and reduce inflammation" if you eat a protein high in the amino acid cystine or "Grow more collagen" if you're consuming a lot of the amino acid glycine.

Fiber

Even though you need to limit your consumption of plant antinutrients, you also need to consume soluble fiber, which is found only in plant foods. Soluble fiber is the kind that can be broken down and consumed by the bacteria in your gut. It feeds beneficial gut bacteria that produce propionic acid and butyric acid, which are known to benefit the digestive tract and the brain.[5] A robust gut microbiome also helps maintain your energy balance and get rid of toxins.[6] Studies show that 20 grams a day of soluble fiber in your diet increases your life expectancy.

One particular soluble fiber, known as modified citrus pectin, has been shown to greatly increase the rate at which the body expels the toxic heavy metals arsenic, lead, and cadmium. In another study, modified citrus pectin also reduced the rate at which cancer spreads in the body.

There is also a second type of fiber, known as insoluble fiber or roughage, that passes through you undigested. The case for roughage is less clear cut, although apparently you need some. Recently I interviewed Robert Lustig, an esteemed endocrinologist at the University of California, San Francisco. He makes a convincing case that roughage functions as a protective barrier in the gut, preventing harmful compounds from reaching the liver.

Carbohydrates

My general rule with carbohydrates is that they are fine so long as you understand exactly what you are getting. Carbs are useful only as an energy delivery system, a way to put glucose into your body. Just remember that they aren't as efficient at producing stable energy for your body as fats are. They are also much less satiating. They tend to produce an energy spike followed by a crash and cravings. If you are going to indulge in carbs, save it for the ones you will truly enjoy.

The real problem with most carbohydrates is the way that nature has packaged them. Many common sources of plant carbs are rich in nutrient-sucking phytic acid and inflammatory lectins. Grains in particular are bad news. Having some carbs in your diet is good, to maintain gut integrity and keep up your thyroid hormone production. But the bottom line is that contrary to what mainstream guidelines suggest, you don't need to eat a diet that is mostly carbohydrate in order to be healthy.

WHERE TO GET THE GOOD STUFF

This is where the biohacking rubber meets the road. A deer can get the nutrients it needs from leaves, fruits, and nuts and then have the energy to shake off a run from a predator. You are not a deer; you need to operate more strategically. When you go to your local market (or your supermarket or your online grocer), you need to know what to buy. Here are some guidelines about where to get the raw materials your body requires.

Grass-Fed Meat

One of my favorite sources of good proteins and good fats is meat from grass-fed animal ruminants. Animal protein is better than protein from plants, partly because it does not contain phytic acid, which inhibits your ability to absorb minerals. Meat also contains creatine, an amino acid that increases muscle mass and energy, protects against neurological disease, and improves brain function.[7] If you eat a lot of meat, you might not notice its benefits because you are getting them all the time. In a 2011 British study, vegetarians who took creatine supplements showed improved memory and faster reaction times. Meat eaters didn't show the same improvements, because they were already operating at a higher level.[8]

Grass-fed beef is a rich source of the important amino acids carnitine, choline, and taurine. It contains vitamin K2, which you cannot get from vegetables, and vitamin B12, which vegetarians are usually deficient in. It also provides you with other B vitamins, such as biotin, which gives you healthy skin and hair, as well as coenzyme Q10, an antioxidant that may provide heart benefits, and absorbable iron and zinc, filling in some of the minerals that you almost certainly need more of.

Beef tallow has been a predominant source of healthy fat in the human diet for a long period of time. When we began eating more vegetable fats, we started becoming a lot less healthy. You may have seen news stories and antimeat campaigns claiming that meat causes cancer. In reality, the opposite appears to be true. A study out of the University of Minnesota found that beef fat can protect against cancer in the gut—in animals, anyway. The same study found that soybean oil may be linked to cancer, by the way.[9]

I'm focusing on grass-fed animal protein for a reason: not because it's trendy but because the quality of the meat is so much higher than that of conventional meat. A grass-fed cow is less fatty than a cow fattened on corn and soy. Its fat also tastes different. To me, it's delicious. (In my house, the kids call the beef fat "candy.") Modern farms aren't even called farms anymore. They're called concentrated animal feeding operations, or CAFOs. These animal factories have changed the composition of beef fat. Fortunately, you have a choice: you can say no to CAFO beef and choose grass-fed beef. Then you will get the good animal fat where all the fat-soluble nutrients are and the right form of stearic acid fat.

Grass-Fed Dairy Products

If you've read my previous books or bought any of my products, you know that I'm a big fan of grass-fed dairy products. I popularized the idea of putting grass-fed butter into coffee, so much so that I think I contributed to a global shortage of grass-fed butter in 2014. Just as with beef, grass-fed butter tastes different from the familiar supermarket kind. It contains higher concentrations of nutrients and more of the right kinds of saturated fats.[10] It also contains a compound called conjugated linoleic acid, or CLA, which can help you lose weight and burn fat.[11]

The milk fat produced by a conventionally raised cow is quite a bit different from milk fat produced by a grass-fed cow. In Canada, a group of dairy farmers recently began feeding leftover palm oil residue to cows, because it was cheap and available. Anything to make a little money, right? Customers started complaining that the butter from those cows wouldn't spread. The palm oil from the feed had made its way into the milk fat, and palm oil is very, very solid.[12] That's how direct a connection there is between the feed that goes into a cow and the milk that comes out.

When a cow is fed grass, that dietary connection does good things for you. Grass-fed cows put out high-quality fats. In addition to CLA and saturated fats, their milk contains much higher levels of healthy omega-3 fatty acids relative to unhealthy omega-6s, according to another study at the University of Minnesota.[13]

Good Plants

Despite the challenges posed by antinutrients and agricultural chemicals, there are large categories of plants that are good and safe to eat. As a general rule, I steer clear of genetically modified foods, or GMOs, in which the genes of a plant (or animal) have been deliberately altered. The most common modified foods are corn, cotton, canola, soy, and sugar beet. Some animal studies have linked GMOs to organ damage, cancer, and liver stress. In 2016, a group of European researchers reported that corn modified to tolerate glyphosate showed signs of oxidative stress pathways within the corn kernels.[14] There is no definitive evidence connecting these problems to human health, but that's a gamble I would not take. The best way to make sure you are getting the natural versions is to choose foods labeled "certified organic" or "non-GMO."

Plants That Are Low in Antinutrients

You want to choose plants that do not trigger inflammation. Some plants are fine: carrots, celery, broccoli, cabbage, lettuce, and asparagus, for instance. If you make a salad, it should look like your grandmother's salad. Iceberg lettuce with some cucumbers? Great. Maybe add some arugula or colored lettuces. Kale, spinach, garbanzo beans, and all sorts of other fancy crap? No, now you're introducing plants that are full of phytic acid. If you want to be fancy, add some fennel. Raw carrots can be really detoxing for the gut, so adding some to your salad is worthwhile.

The most important part of a healthy salad is the dressing, because that's where restaurants and food manufacturers often put bad fat, sugar, and stressful chemicals. Make a simple salad dressing using my biohacker's recipe: ¼ cup of olive oil, ¼ cup of apple cider vinegar, 2 tablespoons of C8 MCT oil (a widely available oil made from medium-chain triglycerides), half an avocado, and some herbs. The vinegar helps with mineral absorption, and the whole thing is delicious.

If you insist on eating kale and spinach, you can lessen the antinutrient burden by steaming them and draining off the water. That way you won't get as large a dose of mineral-sucking antinutrients. You can also limit your exposure to antinutrients in vegetables by removing the peel, which is where lots of plants concentrate their lectins and phytic acid. For instance, cucumbers contain lectins, but most people handle cucumbers just fine if they peel them first. The idea of eating the peel because that's where the vitamins are contained is just wrong; the peel is where all the bad stuff is.

The same is true of grains. White rice is pretty much just a naked grain of starch, but it's better for you than brown rice because it's been stripped of the phytic acid–containing husk. If you want some other grains in your diet, the best one would be sourdough rye bread. Rye contains high amounts of phytase, the enzyme that breaks down phytic acid, and if you sour it properly that should remove most of the plant toxin. Your sourdough will probably still contain some lectins, but in quantities low enough that you can handle them.

This is an important counterpoint to my previous chapter. Look, phytic acid is bad for you, but that doesn't mean you have to be on a zero–phytic acid diet. You just want to keep it below the level where it's going to make you weak. Coffee and chocolate contain a little bit of oxalic acid, another plant antinutrient, but they also have offsetting benefits and pleasures. I also believe in allowing some room for sugar and starch in your diet. Just save your sugar for the things that taste really good, that are good for you, and that you truly enjoy. If you can, choose sweets that are homemade and contain healthy fats (something like your grandma would have made) instead of processed store-bought frankenfoods. Eat the cheesecake, not the Twinkie.

Plants That Are High in Fiber

I told you earlier that you want soluble fiber in your diet, and you aren't going to get that from your meat. Artichokes are a good source; they are low in toxins and are pretty darn good for you. A half-cup serving of artichoke hearts contains 7 grams of fiber.[15] The richest source of soluble fiber—one that is available year-round and that will feed your gut—is acacia gum, derived from the sap of the acacia tree found in Senegal. It's another example of an old native food that is making a comeback now that we understand it can provide soluble fiber without the bad stuff.

Flaxseeds also contain a lot of soluble fiber, but you don't want to eat the full seed because it's high in unstable omega-6 oils. Flaxseed fiber, separated from the rest of the seed, is good for you, however. A surprisingly good source of soluble fiber is the avocado. A typical ripe avocado contains 7 to 10 grams of fiber. It is also rich in potassium and magnesium, along with vitamin E and B vitamins. It does contain some traces of omega-6, but not much. So there's a good reason to eat some guacamole.

You also want to consume some insoluble fiber, as long as you choose it wisely. A common source of insoluble fiber is psyllium, derived from psyllium husk, part of the seed of the *Plantago ovata* plant.[16] The problem with psyllium is that it is really rough on the lining of the gut when it's prepared the way it normally is, for example, in Metamucil. Extremely finely ground psyllium, which you can easily incorporate into your baking, is much gentler and has beneficial effects on the gut. I'm a fan of psyllium, but only if it's very, very finely powdered. You can also get insoluble fiber from celery, fennel, and cruciferous vegetables.

Nutrient-Rich Plants

Speaking of cruciferous vegetables: Brussels sprouts, cabbage, broccoli, and other cruciferous vegetables have been on my recommended list forever. Cabbage is a good source of vitamin K, potassium, and calcium. Brussels sprouts are rich in potassium, calcium, and iron. What I haven't talked about is that you have to be careful not to buy them when they're not fresh, especially in winter. If you feel like crap after you eat cruciferous veggies in winter, it's probably because they were covered with a type of mold called *Alternaria brassicae*,[17] which can give you headaches, make you feel tired, and actually alter your senses. Be especially cautious with out-of-season cabbage and Brussels sprouts.

Asparagus is full of nutrients, including vitamin K and iron, along with a decent amount of soluble fiber. Zucchini and summer squash contain useful doses of potassium and fiber. People generally eat them as vegetables, even though they are technically fruits. The traditional fruits are good for you, too, during summer, when they are in season. Plants want you to eat their fruit because that is part of how they distribute their seeds. As a result, pears, apples, and peaches are solid sources of nutrition. So are many berries (raspberries, blackberries, strawberries, cranberries, blueberries), lemons and limes, and pineapple. The problem is that if you eat sweet fruits in winter, you are going to confuse your body. It's best to avoid too much sugar in winter because sugars are not naturally available during winter and the human body is not adapted to eat them then.

It's also important to eat herbs and spices. Many of them contain antioxidants, along with antifungal and antibacterial compounds.

In particular, coffee, tea, chocolate, and certain spices contain high levels of polyphenols, a family of compounds that have strong antioxidant properties. Polyphenols have almost magical properties: They protect your heart. They reduce the risk of diabetes. They can amplify immunity and restrain cancer. Ignore the common bad advice about "eating a rainbow" of vegetables and eat a rainbow of herbs and spices instead: some turmeric, some oregano, some thyme and rosemary. Even small doses will improve your health.

Most of the herbs that we eat, even vanilla, were originally prized as medicinal ingredients. Now we know that spices and sea salt truly are Mother Nature's sources of essential minerals and health-promoting polyphenols. Salt is often treated as an enemy, but the truth is, most of us don't get enough sodium, so you shouldn't feel guilty about adding it to your food.

| PLEASURE FOODS |

An important theme here is that eating a good diet should not make you miserable. I'll go further and say that if it's making you miserable, it isn't a good diet. Biohacking is about finding shortcuts that will take you to an optimal state of energy and strength so that you can unlock the best version of yourself. Being miserable is not an optimal state.

Chocolate and coffee, which are two of my favorite foods, also contain phytic acid. I'm not going to give up my favorite foods. Instead I carefully manage what I eat and save my phytic acid for the foods that I love the most. Why would I punch myself in the face? It's also okay to have some carbs, whether in the form of starch or even sugar. If you want to have foods containing some honey or 5 or 10 grams of sugar after dinner, you'll be all right.

Sometimes you'll eat food that isn't particularly good for you, but there is a way to hack that: you can take supplemental enzymes to help you break down the bad stuff, as well as supplemental calcium and trace minerals with your coffee to help bind toxins in your gut[18] (and I'll come clean, I am developing a product like that). Supplements are going to be an essential part of your diet no matter what you do, so you might as well adjust your hacks to allow you to eat your favorite foods.

THE SLEEP RESOURCE

People don't usually think of sleep as a resource, but it is definitely something you need to stock up on before you set out to upgrade your MeatOS. When you settle in to a good, deep sleep, your body switches into repair mode. That is when your cells remove their molecular trash and perform maintenance. It is when your muscles grow and your tissues regenerate themselves with the high-quality raw materials you have stocked up on. If you don't give your body a chance to do that, you won't be able to upgrade your MeatOS as efficiently and everything else that you try to do will be harder than it should be. High-quality sleep reduces stress and inflammation, enhances cognitive function and memory, increases your sex drive, and helps you lose weight. In other words, it's a nonnegotiable resource, and it's highly accessible if you know how to do it right. Sleep is also an important part of the recovery process in a later stage of your biohacking. You'll hear more about it later in this book.

4

SUPPLEMENT YOUR MeatOS

A lot of people believe that if you eat a healthy, nutritious diet, you will get all of the vitamins and minerals your body needs. Food ads and public health messages have been driving that idea into our heads for years. Truth is, even if you avoid the friction foods and load up on the good stuff, you still won't have everything you need to optimize your biology and keep your mitochondria running at full power. There's more work to be done before you are ready to start upgrading your MeatOS.

One big reason why you can't get everything you need from food alone is that our modern lifestyle is drastically different from the ancestral lifestyle that we are adapted to. We live most of our lives in an indoor environment, rarely getting enough sunshine to make an adequate supply of vitamin D. We are surrounded by synthetic materials, toxins, and indoor pollutants. We are connected to one another 24/7 through our electronics, engaging our brains and disengaging our bodies in novel ways. We are surrounded by artificial illumination that alters our sleep patterns. No matter how much you may attempt to limit your technology use, there is no way to escape the influence of the modern world, which is placing new demands on your body.

We have also changed the planet we live on. We raise livestock in factories, so we don't have animals eating grass and pooping minerals back onto the soil. We have built dams so that rivers no longer flood and fertilize their banks. Extensive tilling, along with the use of industrial fertilizers, pesticides, and herbicides, has depleted farmland soil of its minerals and complex mix of organic compounds; the ubiquitous herbicide and antibiotic glyphosate, in particular, sterilizes the soil and increases your body's demand for nutrients.[1] Meanwhile we are dumping carbon dioxide into the atmosphere, which is not only changing the climate but also affecting the way crops grow. Multiple studies have found that rising levels of carbon dioxide are reducing the vitamin content of food crops and cutting their mineral concentrations by up to 15 percent.[2]

Meanwhile, here you are, not just trying to stay normal but reaching for an upgrade of your MeatOS. But you won't get as much value from exercise, meditation, sleeping, or any fancy hacks if your body is missing the foundational vitamins and minerals.

This chapter and the next will teach you how to think about using supplements to make sure your body has the building blocks it needs to help you hit new levels of performance. If you get this right and you do nothing else in this book, you're going to have a much better life than most people do. Conversely, if you simply jump ahead to the chapter on building muscle or having a better brain and you diligently do the things you learn there, your results might disappoint you. Simply put, don't skip this chapter. The return on investment for consuming proper minerals and basic vitamins is higher than you might think, and the risks of supplementing are very low. Supplements are profoundly safer than drugs. In the past twenty-seven years, there have been zero recorded deaths from vitamins but about 3 million deaths from pharmaceuticals.[3]

There is a lot to take in here. If you don't want to know the details of why my recommended supplements work, you can skip to the end of each section to get to the "TL;DR" (too long; didn't read) recommendations. I'll also tell you about a few important herbal supplements that can help you get to your goal faster or with less work. Let's get started.

| VITAMIN SUPPLEMENTS |

The four essential fat-soluble vitamins I singled out in the last chapter—D, A, K, and E—work synergistically with one another. A lot of nutrition research has focused on each member of the four-some individually, ignoring the fact that they interact. You are probably deficient in all of them, but you will get into trouble if you supplement some of them without tending to the others. It's like listening only to the violin part to figure out why a symphony is beautiful. You're going to need all four vitamins at the right levels, working together. D and A, in particular, work together in harmony. It doesn't make sense to take them separately except in special cases. That is why I like to lump the four of them together as a single nutritional unit, vitamin DAKE.

When you scrape away all of the hype and the marketing, vitamins D, A, K, and E are the most foundational of all the vitamins. Your MeatOS needs minerals to do its job, and the fat-soluble vitamins are the cargo vessels that get the minerals where they need to go. Vitamin DAKE is especially critical now that most people are deficient in many essential minerals. Without those vitamins, your body can't fold proteins to build cells; it can't move electricity to carry messages; and it can't build enzymes to generate energy in your mitochondria. Start with the vitamins so you can use your minerals well, then get the minerals so that your body can function properly. Every other type of supplement is secondary to this two-part foundation.

Even with a careful diet, you're unlikely to get enough of vitamin DAKE in the right ratios from your food. If you are mostly plant based or vegan, you have no chance. If you eat large amounts of liver and organ meats every day, you can get most of these fat-soluble vitamins, but even then you should supplement to get your dose just right.

The good news about fat-soluble vitamins is that they are stored in your body's fat reserves, so it takes a while for them to be depleted.

The bad news is that it can also take a while to get your levels up to where you want them. It is easier to overdose on fat-soluble vitamins than it is on water-soluble vitamins. The risks are still fairly low, but you do have to be a little more careful with them.

Here you are going to learn everything you need to know about the four fat-soluble vitamins so you don't have to think about them ever again if you take a supplement at least some of the time. It doesn't have to be hard.

Supplement Your Vitamin D

Mother Nature did her best to make sure we wouldn't have to worry about not having enough vitamin D. Your body will manufacture it out of cholesterol if you expose enough of your skin to enough sunshine and you have the right vitamin D receptor settings in your genes. The problem is that our modern lifestyle prevents that from happening: unless you live in a sunny part of the world and spend most of your time outdoors naked, as our ancestors did, your body can't make an adequate supply. Sitting under office lights, being lit up by your laptop, is no substitute.

Maintaining the proper level of vitamin D reduces your chances of getting an infection or having other immune problems.[4] It also helps you maintain a healthy blood pressure, reduces asthma and allergies, and can reduce the risk of developing breast, prostate, colorectal, and ovarian cancers.[5] It even helps with sleep (although only if you take it in the morning). Oh, and it helps keep your metabolism working efficiently and may help reduce the likelihood of developing diabetes or depression. It helps both men and women maintain healthy hormone levels. It's also good for preventing osteoporosis.[6]

Vitamin D has a dark side, though, because it raises the calcium level in your blood. When your calcium level gets too high, it can cause problems. The most common problem is kidney stones. A high calcium level also contributes to plaque formation in your arteries and heart disease. You're likely taking vitamin D to lower cardiovascular risk, but if you're taking it without its partners, A, K, and E, you're increasing the risk of developing calcified plaque. The government's recommendation for vitamin D is only about 400 IU per day because it has no way of knowing whether you are taking other stuff along with it.

The reality is that taking too much vitamin D or too much calcium by itself can cause calcification in your body where you don't want it; so can eating too much processed food that contains lots of (unlabeled) phosphorus. Vitamin D's toxicity seems to be caused by its depleting vitamin K.[7] That's why you need to pair it with A, K, and E. That's also why the second half of this chapter shows you how to upgrade your mineral levels. After all, supplementing with vitamin D won't work if you're magnesium deficient.

There are two forms of vitamin D, known as D2 and D3. Big Food likes to use vitamin D2 (ergocalciferol), which is available in fortified foods and many supplements, and pretend that it's the same thing as the real, more bioavailable vitamin D3 (cholecalciferol). D3 is the type of vitamin D that your skin makes from cholesterol when you go out into the sun. You can get D3 in some animal foods, too. Even if you eat a lot of egg yolks from pastured chickens or consume wild-caught salmon, you're probably not going to get enough D3 from food. Cod liver oil is the exception. Unless you are interested in routinely sunbathing in the nude or love chugging cod liver oil, you might want to supplement.

An average person requires about 1,000 IU of D3 per twenty-five pounds of body weight per day. That means a baseline for most of us is somewhere around 5,000 IU per day. However, people with weak vitamin D receptors in their genes (like me) or heavy people (like me) may need a lot more in order to get their blood levels high enough for high performance and lower risks. Because everyone responds differently to vitamin D supplementation, it is important to get your level tested routinely when you're taking it. The nonprofit Vitamin D Council[8] and most of the doctors I work with recommend blood levels of 25-hydroxyvitamin D, or 25(OH)D—the parameter used to assess vitamin D status—of 70 to 90 nanograms per milliliter (ng/mL).

The best thing you can do is take a combined DAKE supplement, then add additional vitamin D3 taken in the morning until you achieve blood levels of 70 to 90 ng/mL using a cheap blood test you can do at home. If you eat a very-high-calcium diet, you can get away with a lower vitamin D level, as low as 50 ng/mL. If you eat a lot of junk food and drink soda, you'll need even more vitamin D3. If your body stores a lot of fat, the D3 you take will be stored there, so you won't be able to use it; you'll have to increase your D3 intake until you lose the extra fat!

TL;DR FOR VITAMIN D3

Take 5,000 IU, more if you're heavy or have dark skin, less if you eat lots of calcium. Never take it without vitamins A, K, and E.

Supplement Your Vitamin A

Big Food has tried to convince you that plants are a good source of vitamin A, which is false. Real vitamin A, called retinol, comes only from animal foods. Plants contain carotenoids (marketed as provitamin A), which your body can potentially turn into real vitamin A—but only by expending a lot of energy, provided you eat enough animal protein, have enough vitamin E and enough minerals in your body, and don't eat a lot of fiber. It won't work if you are hypothyroid or have high levels of heavy metals in your body. Getting enough vitamin A from the beta-carotene in plants is a vegan fantasy that won't help you thrive, although adding butter to colored veggies will at least help you absorb the carotenoids better. But even a diet full of animal-based foods is likely not enough. You would have to eat two to four egg yolks per day, and a lot of butter or full-fat dairy products to get enough retinol from food unless you eat about an ounce of liver per day or take cod liver oil. Let's face it, you're probably not going to do those things reliably if you ever travel or eat at restaurants.

Fortunately, taking a vitamin A supplements is easy. There are two reasons to take vitamin A as part of the full DAKE set. The first is that vitamin E helps your body absorb vitamin A. The second is that vitamin A seems to protect you from the downside of taking higher doses of vitamin D, which is one more reason to take them together.

You need a full supply of the real vitamin A. It improves night vision, dry eyes, immune function, and sleep. It helps maintain your circadian rhythm, and, like vitamin D, it can reduce autoimmunity and improve allergies and hormone levels. Vitamin A helps maintain your mucus membranes, which is why your lungs and GI tract rely on it to run at full power. Your pancreas also needs vitamin A in order to make insulin. Studies show that Vitamin A depletion can cause mitochondrial dysfunction—a lack of performance in your cellular powerhouses[9]—leaving you feeling tired and lazy. Just be mindful not to overdose. Because it is fat soluble, you can get too much of it. Excess vitamin A can give you itchy skin and cracked lips, weak bones, and hair loss.

As with the other fat solubles, vitamin A interacts with minerals. It helps your body absorb iodine so your thyroid works, and your body needs zinc in order to move vitamin A around in the body. If you are trying to get vitamin A from plants, you're going to need extra iron. For most people, I recommend 5,000 to 10,000 IU of vitamin A taken with a meal (or coffee) that contains fat, less if you eat liver regularly. In 2020, the measurement units for vitamin A in the United States changed to "mcg RAE" to more easily enable Big Food to claim that plant vitamin A is the same as real vitamin A; 1,515 mcg RAE of retinol is the same as 5,000 IU.

TL;DR FOR VITAMIN A

5,000-10,000 IU per day of preformed retinol (not beta-carotene), always taken with vitamins D, K, and E.

Supplement Your Vitamin K

Vitamin K never seems to get the same love that the other fat-soluble vitamins get, even though it is just as important. Whereas vitamin

D helps you absorb calcium, vitamin K helps keep it in your bones, where it belongs. Vitamin K also keeps calcium from accumulating in your kidneys so you don't get calcium-based kidney stones, and it keeps calcium from building up in your blood vessels, where calcified plaque is particularly dangerous. Likewise, calcified arthritic joints are not good for performance, and vitamin K helps prevent calcification.

You need vitamin K to have healthy blood clotting, but even a high level of vitamin K does not increase the risk of excessive clotting or stroke, unless you are taking a prescription medication such as warfarin. Since warfarin induces a vitamin K shortage, I don't recommend it as a sustainable long-term treatment for blood thinning. Talk to your doctor before taking vitamin K if you are on warfarin. Perhaps most impressively, vitamin K can have a profound impact on reducing or even healing cavities in teeth. Like the other members of DAKE, it helps your body maintain sex hormone levels and may reduce cancer risk. It also improves blood sugar control, which can be a game changer for achieving stable energy levels and avoiding postmeal sluggishness.

There are several different kinds of vitamin K,[10] but supplements usually provide vitamin K1 or K2 in one of two forms, called MK-4 and MK-7. Each does different things. K1, found in plant sources such as kale, helps your blood clot. The real superhero is vitamin K2, which you can find in animal foods such as butter. There are two subsets of vitamin K2, MK-4 and MK-7, both of which are important. Vitamin K2 works with vitamin D and calcium to keep calcium where it needs to be: in your bones instead of in your arteries. It can also help prevent osteoporosis, atherosclerosis, cancer, and inflammatory diseases.

I allow the animals on my farm to eat a lot of leafy greens so they convert the plant vitamin K1 into vitamin K2, and I take supplements containing both K1 and K2. Vegans can eat a fermented soy product called natto, which provides vitamin K2 in various forms; otherwise they will be deficient. Vitamin K2's MK-4 form is the most important form of vitamin K, followed by MK-7. Aim for anywhere from 200 to 2,000 mcg as a part of DAKE. Based on researcher Chris Masterjohn's epic K2 database, you can get 200 mcg from 7 grams of natto (slimy!), 64 grams of beef liver (enough to raise your purine level, which is not good), five egg yolks, about 100 grams of cheese, 300 grams of grass-fed butter or lard, or 110 grams of dark-meat chicken.[11] Some exotic foods, such as emu oil, duck fat, and goose liver, are also very high in K2. The reality is that almost no one eats much of those foods consistently, which is why supplements are the best way to get your K2.

K2 is simply not optional if you take vitamin D3, and vitamin D3 is not optional if you want to perform well. Unlike the other members of DAKE, though, you could take vitamin K2 by itself and still benefit. It is hard to get enough vitamin K for it to become toxic, so you don't have to worry about overdosing.

TL;DR FOR VITAMIN K

2,000 mcg of vitamin K containing 1,000 mcg of K1 and 1,000 mcg of K2 as MK-4 and possibly also MK-7 per day.

Supplement Your Vitamin E

Your body's laziness principle kicks in if you are deficient in crucial nutrients. About 74 percent of Americans are deficient in vitamin E, according[12] to the Food and Drug Administration. Time to get supplementing.

Vitamin E has a weird reputation. On the one hand, Big Food loves to talk about vitamin E; on the other, it doesn't tell you that it is using a synthetic form that doesn't work very well and is not found in nature. The synthetic form you should avoid is called DL-alpha-tocopherol. The reason it's a problem is that it contains a mixture of mirror-image molecules, or *stereoisomers*, that can cause negative health outcomes. Big Pharma loves to talk about vitamin E because it can do short-term studies with the synthetic form to "prove" that vitamin E supplementation can be bad for you. Real vitamin E is not bad for you and has very positive effects in your body. It acts as an antioxidant and can protect polyunsaturated fats in your cells from oxidation. This is important because too much oxidation can lead to reactive oxygen species that damage your mitochondria and sap your energy.

Vitamin E makes your brain work better and reduces your chances of developing heart disease, Alzheimer's disease, and cancer.[13] Studies show that people with higher vitamin E intake have better cognitive performance.[14] When your brain works better, your whole body has more energy to perform. Like the other vitamins in DAKE, it helps with hormone balance and immunity. If you follow my food recommendations and eat grass-fed butter and meat and minimize your intake of inflammatory omega-6 oils, you will need less vitamin E. If you have been eating soybean oil, canola oil, corn oil, and anything fried at a restaurant, you want to take extra vitamin E for four to six years. That's because it takes about two years to replace half the fat in your body with better fats. If you stop eating bad omega-6s now, in only six years, about 87.5 percent of your cells will have replaced the bad omega-6 fats with better ones. Then you'll need less vitamin E because you'll have less inflammation. Your body mostly stores vitamin E in fat, but there are also extra-large amounts in your uterus or testes, depending on how you are equipped.

There are eight forms of vitamin E, and each has varying levels of effectiveness. You will find alpha, beta, gamma, and delta forms of tocopherol-type vitamin E in common supplements, but new research points out that you will benefit differently from taking a form of vitamin E known as tocotrienols. Tocotrienols are unsaturated, so they fit into cells differently. The most powerful tocotrienols come from the annatto tree. This unusual tree is known as "the lipstick tree" because it is so rich in colored compounds that impact health. The vitamin E found in annatto is mostly delta- and gamma-tocotrienols. It has unique benefits for helping reduce the buildup of fat in the liver, protecting cells, reducing chronic inflammation, and improving eye and bone health.

The annatto tree also contains geranylgeraniol, a vital component of cells that declines with age. Geranylgeraniol is necessary for building cell membranes, creating energy in mitochondria, and making sex hormones such as testosterone. It also shares some superpowers with vitamin K because it helps keep calcium out of places where it doesn't belong. This little-known antiaging molecule has a lot of promise. Since it comes with vitamin E from annatto, I recommend taking both together even though geranylgeraniol is not technically vitamin E. The best option is to take 150 mg of mixed delta- and gamma-tocotrienols, along with 150 mg or more of geranylgeraniol, and to eat plant and animal foods that contain common tocopherol forms of vitamin E. Take more if you insist on eating fried foods or seed oils.

TL;DR FOR VITAMIN E

150 mg of delta- and gamma-tocotrienols with 150 mg of geranylgeraniol. Don't eat seed oils.

TL;DR FOR VITAMIN DAKE

Vitamin D3: 5,000 IU per day, more if your skin is dark or you are heavy or until your vitamin D blood tests show a level of 70–90 ng/mL.

Vitamin A: 5,000–10,000 IU per day as preformed retinol (not beta-carotene) (same as 1,515 mcg RAE).

Vitamin K: 2,000 mcg containing 1,000 mcg of K1 and 1,000 mcg of K2 as MK-4 and possibly also MK-7 per day.

Vitamin E: 150 mg of delta- and gamma-tocotrienols with 150 mg of geranylgeraniol.

| HERB AND SPICE SUPPLEMENTS |

Modern biohacking sometimes leads us straight back to the traditional practices and wisdom of cultures around the world. The original spice trade across the continents didn't just give people access to intense flavors, it also provided a way for them to get their hands on the original mineral and vitamin supplements. For example, there is a meaningful amount of manganese, a trace mineral, in saffron, turmeric, and cinnamon.

The herbs on my go-to list are high in antioxidants and are great at reducing inflammation to ideal levels, which is key to achieving optimal functioning, enhanced immunity, mental clarity, and all the other things that give you the kind of energy you want to have all day long. Although cooking with herbs and spices is great, supplements are sometimes more easily absorbed by your body and can be more convenient. You want to aim for around 1 or 2 grams of each per day (follow manufacturer's guidelines) for enhanced performance.

Turmeric.[15] Its benefits to your MeatOS are largely pinned to curcumin, a potent antioxidant that dramatically reduces inflammation, while also giving the spice its vibrant yellow color. According to a 2018 UCLA study, a daily dose of curcumin improves memory and mood in people with age-related memory loss.[16]

Pair a curcumin supplement with bromelain (a digestive enzyme found in pineapple), and avoid the many popular supplements that contain black pepper, which has been linked to leaky gut syndrome.[17] Include turmeric in your diet by adding it to salad dressings and meat or fish marinades or by making a turmeric-infused tea or latte. If you're supplementing, you'll want to take about 1 to 2 grams of turmeric extract containing 95 percent curcuminoids per day.

Panax ginseng. Traditional Chinese medicine uses ginseng for a wide variety of treatments, including as a performance enhancer and immune booster. You may have also heard that ginseng enhances libido, although that aspect is overhyped. There are more than a dozen forms of *Panax ginseng.* Only five of them are used medicinally; two very popular ones are Korean red ginseng and white ginseng.

Studies show that ginseng is effective when used to improve cognition and focus.[18] The improvement in cognition is most likely the result of a decrease in fatigue. In studies where individuals were not already experiencing fatigue, they did not see an increase in cognition. Other research has shown that in healthy people, ginseng can have a mood-elevating effect and can increase calmness and improve memory and performance.[19] Just be sure to consult with your doctor before taking ginseng if you are taking any medications, as ginseng can decrease the effectiveness of certain medications.

You can take *Panax ginseng* as a supplement or add it to your tea. You can also use it as a root. Aim for 1 gram per day for enhanced performance.

Siberian ginseng, aka eleuthero. Traditional healers have used Siberian ginseng to fight fatigue, maximize physical performance, and improve overall immunity and longevity. Research backs their practices. Looking to increase your endurance? In one study, Siberian ginseng increased subjects' time to exhaustion by more than 500 percent.[20] Ginseng can improve resistance to both cognitive and physical fatigue.[21] There's also promising evidence that this adaptogen has immunity-boosting effects and can increase T cell count.[22]

Siberian ginseng is best used as a capsule or tincture. As a capsule, take 100 to 200 mg twice per day or as directed. As a tincture, use it as directed on the package.

Cinnamon. Cinnamon is best known as a blood sugar–stabilizing superhero, but it also contains a host of antioxidant, anti-inflammatory, and antibacterial compounds, making it a powerful tool for preventing systemic inflammation.[23]

Cinnamon's claim to fame is its ability to lower blood sugar in diabetics by activating insulin receptors.[24] It also contains a host of compounds with antioxidant and anti-inflammatory properties that can lessen the likelihood of cellular damage and chronic disease. Cinnamon inhibits the NF-κB proteins, transcription factors for pro-inflammatory genes and genes involved in immune, growth, and cell death responses, and it prevents blood platelets from clumping—all of which protect against heart disease and other diseases of inflammation.[25] Cinnamon also blocks growth factors associated with abnormal cell growth, possibly protecting against cancer.

There are two common types of cinnamon that you'll find in the US market. The first is Ceylon, or "true," cinnamon, which comes from the Ceylon cinnamon tree, an evergreen native to Sri Lanka and parts of India.[26] This is the best cinnamon for your body. The cheaper, more popular cassia cinnamon contains high levels of coumarin that can damage your liver and harm your performance.[27]

Inflammation is the root of many chronic and age-related diseases, and the anti-inflammatory compounds in many herbs and spices can help you minimize damage to your cells. Cinnamon is known to down-regulate inflammatory cytokines and increase the production of anti-inflammatory proteins. Additionally, compounds from cinnamon extracts are potent antioxidants and can hunt down free radicals associated with chronic inflammation. You can add cinnamon to smoothies or sprinkle it into your coffee. For most people, 1 to 2 grams of Ceylon cinnamon each day is safe.

Saffron. Healers have been using saffron for thousands of years for digestion and detox and even to treat tumors. Recently, saffron and its active component, crocin, have caught the attention of researchers, who are studying the compound for its protective effects on the brain, its ability to balance mood, its potential to help you lose weight, and more.

Saffron comes from the *Crocus sativus* flower, a relative of the iris. Workers painstakingly hand-harvest the dark red filament in the pistil and dry it to make the spice. Each flower yields just three saffron threads, which explains why saffron is so expensive. Researchers have shown that crocin works in the brain by keeping your neurons young and sharp, possibly through its antioxidant action.[28] Its snapping up of free radicals helps keep the immune response in working order, which protects healthy cells from attack.[29] It also reduces inflammation in the brain and inhibits certain markers of Parkinson's disease, a neurodegenerative disorder.

You can use saffron in cooking, but you're not likely to get the research-backed effective dose just by eating a bowl of bright yellow rice. If you want the mood, brain, and weight loss benefits, look for saffron in a liquid extract or in capsules. You'll want to take 30 to 100 mg of saffron extract daily.

Holy basil. Studies show that this herb, also known as tulsi, can be an effective liver protector,[30] especially when paired with milk thistle. It has been shown to function as a stress reducer, antioxidant, and antianxiety supplement[31] and appears to promote longevity as well. As a bonus, holy basil is known as both an antifertility agent in men and a libido enhancer, so this could be a fun herb for couples not ready to conceive.

You can use holy basil as a garnish on your dishes, add it to your water, or use it to make a tea, to provide you with at least 1 gram daily. You can also purchase a powder or supplement form. I love the anti-inflammatory effects of holy basil, but I have some concern that anything that reduces fertility may be harmful to the rest of your cells, too. The health of your little swimmers (or your eggs) is a great sign of how healthy your overall system is. But I take it whenever I'm inflamed, because it works.

Rhodiola rosea. This herb, also known as arctic root or golden root, can aid your biohacking by keeping fatigue and exhaustion at bay. This potent root contains more than 140 active ingredients, the most bioactive ones being rosavin and salidroside. Rhodiola is used in traditional Chinese medicine and is often used to promote vitality and immunity. It can reduce fatigue and exhaustion in prolonged stressful situations and can also reduce C-reactive protein levels.[32]

There's good evidence that rhodiola improves cognition, independent of fatigue reduction. A study on the effects of rhodiola on night duty–caused fatigue showed that the herb can improve performance by about 20 percent, regardless of fatigue level.[33] It might also improve your mood and decrease the symptoms of depression, perhaps because of its effect on serotonin levels.[34] Rhodiola is best taken as a supplement or tea. Too much or too little won't give you the benefits you desire. The sweet spot seems to be about 250 to 500 mg of rhodiola extract, standardized to 3 percent rosavin and 1 percent salidroside, daily.

Sage. This herb derives its primary health benefits (and its flavor) from carnosic acid and carnosol, two anti-inflammatory molecules. It may protect against inflammation-based neurological conditions such as Alzheimer's disease,[35] improve memory and concentration, and reduce anxiety. Carnosic acid and carnosol can also have antioxidant and anticancer effects.[36] Camphor, another sage constituent, kills bacteria and fungi, and still other sage-derived compounds are effective antivirals.

In the kitchen, sage goes well with winter squash, in sausages, and with meat roasts. Consume about 1 to 3 grams of dried sage daily for the greatest benefit. You can also supplement with 1 to 2 grams of sage leaf extract.

Rosemary. This herb contains some of the same antioxidant and anti-inflammatory compounds as sage, along with another that is appropriately named *rosmarinic acid*. Both rosemary and sage increase the activity of superoxide dismutase, an enzyme that removes the free radicals (electrically charged molecules) associated with chronic inflammation.[37] The cooked herb produces the greatest activity, so use rosemary to flavor roasted vegetables, meats, or other cooked dishes: another flavorful biohack. You can benefit from uncooked rosemary, too, as its components include apigenin, a compound that can inhibit the growth of pancreatic cancer cells, and diosmin, which helps prevent hemorrhoids.

If you're going to cook something in oil (sauté, fry, or broil), add some rosemary to the oil to allow its antioxidants to help preserve the oil from oxidation. I recommend about 2 grams of dried rosemary daily.

Bacopa monnieri. This small water plant, also known as water hyssop, a native of India, is an adaptogen: it helps your body adapt to stress. It also improves memory in healthy adults[38] and enhances attention and mood in people over age sixty-five.[39] Scientists still don't fully understand how it works, but they do know it takes time to kick in; the study participants didn't feel its memory-enhancing effects until they'd been supplementing with it daily for four weeks. If you try bacopa, stick with it for a month before you give up on it.

You want to take at least 750 mg daily. Take bacopa with a fat source to increase its absorption.

Ashwagandha. This is one of the most common adaptogens, used in Ayurvedic practices to promote a state of calm. (Remember that stress can come from exercise, poor diet, infection, feelings of fear, and even your mother-in-law. Your MeatOS doesn't care about the source, only how you deal with it.) Several studies have shown that ashwagandha decreases anxiety,[40] stress,[41] C-reactive protein levels, and cortisol levels. The decrease in cortisol is worth talking about, especially when you compare the effects of ashwagandha to those of other stress-reducing supplements. Studies show that ashwagandha decreases stress by 14.5 to 27.9 percent in healthy but stressed people.

Another useful aspect of ashwagandha is that it shows promise in improving memory formation. This could be important in research for treating Alzheimer's patients.[42] More large human studies are needed to show how and why it might be effective, but there is research to suggest that ashwagandha could reverse the effects of neurological toxins associated with neurodegenerative diseases.[43]

You can take ashwagandha as a powder or in supplement form. You can also get creative in the kitchen and try out some adaptogenic recipes, such as a stress-reducing drink or snack. As a powder, take 3 to 6 grams of powder per day. As a pill supplement, take one 300 mg capsule two to three times per day.

TL;DR FOR HERBS AND SPICES

Aim for 1–2 grams most days of each herb and spice in dried, capsule, or tincture form.

Turmeric: 1–2 grams of turmeric extract containing 95 percent curcuminoids taken with bromelain and a fat source. Avoid supplements containing black pepper extract.

Panax ginseng: 1 gram per day.

Siberian ginseng: 100–200 mg twice per day or as directed.

Cinnamon: 1–2 grams of Ceylon cinnamon per day.

Saffron: 30–100 mg of saffron extract per day.

Holy basil: 1 gram or more per day.

Rhodiola rosea: 250–500 mg of rhodiola extract, standardized to 3 percent rosavin and 1 percent salidroside, per day.

Sage: 1–3 grams of dried sage or 1–2 grams of sage leaf extract per day.

Rosemary: 2 grams of dried rosemary per day.

Bacopa monnieri: 750 mg per day along with a fat source.

Ashwagandha: 3-6 grams of powder per day, or one 300 mg capsule 2 to 3 times per day.

PREBIOTICS, PROBIOTICS, AND POSTBIOTICS

Your gut contains an entire ecosystem of bacteria that digest food, produce chemicals, and participate in your overall biological operation. Think of them as the distributed MeatOS, running on a separate machine but all part of the same network. While you are working to improve yourself, you want to improve the beneficial gut microbes, too.

Probiotics can be tricky, because what works well for one person's gut may not work for another person's and not all probiotics are created equal. Some of them, such as the ones in yogurt, can prompt the production of histamine, leading to bloating and brain fog. There are also some probiotics that degrade histamine, and these are the ones I recommend using: *Bifidobacterium infantis*, *Bifidobacterium longum*, *Lactobacillus plantarum*, and soil-based organisms. Follow the dosage recommendations on the bottle.

Prebiotics feed the helpful bacteria that live in your gut. You can get prebiotics from vegetables, blueberries, coffee, or cooked and cooled starch, or you can supplement with prebiotic fiber. Acacia gum is a great source of prebiotic fiber and you can find this as a powder at health food stores. Your gut may need some time to adjust to prebiotic fiber, so start with a small dose of about 5 grams per day, working your way up to 15 to 20 grams per day.

Postbiotics are a fairly new concept, but they are essentially the helpful by-products that your gut bacteria make. You can supplement what the bacteria do to make sure you have enough of them. Butyric acid, or butyrate, is one of my favorites. This short-chain fatty acid has been shown in studies to reduce inflammation[44] and improve brain health.[45] You can find it in pill form, providing about 1 gram per day, or you can just eat a lot of butter, which naturally contains butyrate. Urolithin A, another bacterial by-product, slows down aging and significantly increases muscle output.[46] Unfortunately, you need a rare combination of gut bacteria to get enough. Fortunately, this compound is available as a supplement in powder and pill form. The clinically effective dose is 500 to 1,000 mg per day.

TL;DR FOR PROBIOTICS, PREBIOTICS, AND POSTBIOTICS

Probiotics: Choose the histamine-degrading probiotics *Bifidobacterium infantis*, *Bifidobacterium longum*, *Lactobacillus plantarum*, and soil-based organisms. Follow the manufacturer's dosing instructions.

Prebiotics: Work up to 15–20 grams per day from vegetables or acacia fiber powder.

Postbiotics: To reduce inflammation, take 1 gram of sodium butyrate per day. For increased muscle output and slower aging, take 500–1,000 mg of urolithin A per day.

| ENERGY SUPPLEMENTS |

Every essential resource in your body ultimately aids your metabolism and furthers your agenda of redirecting the laziness principle in your biology. But some supplements aim directly at ground zero of your energy supply, the mitochondria.

PQQ, or pyrroloquinoline quinone, is an antioxidant that protects your cells from free radicals. Studies show that PQQ can increase mitochondrial density,[47] reduce inflammation,[48] and protect your heart.[49] In humans, the regular form of PQQ turns into rocks when it is exposed to stomach acid. I recommend using a form called Active PQQ that protects it from the action of your stomach acid. Look for a supplement that contains 10 to 20 mg.

CoQ10, or coenzyme Q10, works well with PQQ, helping your body absorb it. Like PQQ, CoQ10 is an antioxidant and is necessary for cell communication, mitochondrial function, and ATP production. It also may enhance blood flow and protect blood vessels.[50] Typical doses range from 30 to 150 mg per day.

Oxaloacetate is an intermediary in the crucial set of energy-generating chemical reactions known as the Krebs cycle. Studies show that it activates mitochondrial biogenesis in the brain, reduces inflammation, and stimulates neurogenesis, or the growth of new nerve cells.[51] Most supplements provide 100 to 200 mg.

Acetyl-L-carnitine is an amino acid derivative that plays an important role in transporting fatty acids across the mitochondrial membrane so your mitochondria can use them for energy.[52] It may be a useful supplement to help prevent or manage diseases associated with mitochondrial dysfunction such as insulin resistance or chronic heart failure. 1 to 3 grams per day seems to provide the most benefits.

TL;DR FOR ENERGY SUPPLEMENTS

PCQ: 10–20 mg of Active PQQ per day.

CoQ10: 30–150 mg per day.

Oxaloacetate: 100–200 mg per day.

Acetyl-L-carnitine: 1–3 grams per day.

| AMINO ACIDS AND PEPTIDES |

Earlier, I mentioned that proteins are constructed of smaller molecular building blocks called amino acids and that nine of the twenty naturally occurring amino acids are essential to your diet because your body can't manufacture them on its own.[53] The essential amino acids are histidine, isoleucine, leucine, lysine, methionine, phenylalanine, threonine, tryptophan, and valine. They are responsible for a variety of processes in the body, including building muscle, making hormones, and repairing tissues. You can get these amino acids from protein-rich foods such as grass-fed beef and eggs, but the ability of your body to absorb and use them varies depending on a wide variety of factors. And sometimes you might not feel like eating a steak after your workout.

Supplementing with essential amino acids gives your body all of the raw materials it needs for growth and repair, in the correct ratios. Studies show that essential amino acid supplements help build muscle and enhance fat loss.[54] They are also extremely usable by the body and produce significantly less nitrogenous waste than other protein sources do.[55] You can find essential amino acids in a powder form that is easy to mix into water. I recommend taking 10 grams before bed or after a workout, apart from other protein sources.

When you string together a sequence of amino acids, you get a peptide. Specific peptides tell your body to carry out specific processes. Trinity signaling peptides are a particularly useful combination of three clinically studied bioactives that tell your body to increase collagen production in targeted places: your skin, your joints, and your bones. This leads to fewer wrinkles,[56] stronger joints,[57] and improved bone density.[58] Taking them is much more effective than taking collagen supplements because instead of being building blocks for collagen, they are molecules that signal the body to build more collagen in the most important areas. I recommend 10–20 grams per day to get the most benefits.

TL;DR FOR AMINO ACIDS AND PEPTIDES

Amino acids: 10 grams of essential amino acids during a workout or before bed.

Peptides: 10–20 grams of bioactive collagen peptides per day.

| TIPS FOR SUPPLEMENTING |

* None of these supplements operates alone. That's Western thinking. If you are deficient in one, there are reverberations in the rest. You need to get the ratios right.

- When buying supplements, look for trustworthy brands carried by major retailers, with clear and complete labeling.
- Beware of "vanity ingredients" that are present in doses far too small to be useful. You want to avoid such products. Refer to the doses above.
- My routine: In order to get my trace minerals every day, I drink Danger Coffee, which has trace minerals built in. I take 75 to 150 pills a day, but I'm an extreme case! My daily dose includes fat solubles, minerals, cognitive enhancers, and herbals plus a few function-specific things, including for sleep and for sex.
- Everybody's needs are slightly different. When in doubt, contact the supplement manufacturer. Follow the guidelines, but experiment. You do you.

5

CHARGE UP WITH MINERALS

As important as vitamins are, they are just the warm-up act. Minerals are the biochemical superstars as you prep your MeatOS for a big upgrade. Yet minerals are probably the least sexy group of supplements, often treated as an afterthought that follows herbals and probiotics. Minerals are also widely misunderstood. Most people know that bones contain calcium. Some may recall that bones contain magnesium as well. You probably know that blood has iron in it. But that's just the beginning of the story. Few people realize that there are more than twenty minerals in their bones alone and they all function in tandem. You can take all the calcium pills you want and still have osteoporosis if you're missing those other critical minerals.

Quite literally, you would not be alive without minerals. Your body carefully allocates dozens of minerals throughout your tissues to build cellular components and—even more important—to activate enzymes. Enzymes are the chemical lubricants of the engine of life. They enable chemical reactions to take place using far less energy than they would otherwise, and they are the basis of biochemistry as opposed to physical chemistry. Enzymes work by binding to a substrate, or chemical input, and bringing together molecules so that they can react with each other. All of the biochemistry that drives your metabolism depends on enzymes. With them, your body runs smoothly. Without them it won't run at all—full stop.

You need your enzymes, and your enzymes need chemical helpers called *cofactors* to activate and go to work in your body. Minerals are the cofactors of many life-critical operations. And you almost definitely aren't getting enough of them. As with vitamins, even a healthy, well-chosen diet will leave you mineral deficient.

If you don't have enough minerals, your body will conserve them for the most critical processes to keep you alive, but that means that other processes that keep you healthy or promote longevity will be sacrificed. We have reached a crisis level of mineral deficiency, and we did it to ourselves with our modern agricultural practices. Even on organic farms, we face a nutrition problem because we keep growing the same plants in the same soil, and rivers don't flood the soil to deposit mineral-containing silt the way they did in the days before large-scale irrigation and land engineering.

Farms that follow the principles of regenerative agriculture, treating the soil as an ecosystem that needs to be nurtured and sustained, use plant matter and animal manure to keep the soil mineralized. Food from regenerative farms is usually high in minerals, closer to the way it used to be. On my small thirty-two-acre regenerative farm, our plants grow in soil that came from the bottom of an organic pond. Using this high-mineral soil on our crops enabled us to reach full productivity in two years instead of a typical four years, and we have impressive yields. More important, the food tastes amazing because our soil has adequate minerals for the plants to take up. But most people don't have access to such farms or such nutrient-rich foods.

Soil depletion is only part of the reason why you need to take mineral supplements. Changes in industrial food production over the last hundred years have created foods that literally suck minerals out of your bones and your body. Our ancestors knew about this problem. When they ate grains, they soaked them, they sprouted them, they processed them, sometimes for several days, or they fermented them. They didn't do it for flavor; they did it to get rid of the plant toxins that steal your minerals, recognizing through trial and error that processing made the plants healthier to eat.

Big Agriculture and Big Food have abandoned traditional practices and inundated the public with phytic acid,[1] the sinister antinutrient I discussed in chapter 2. You find it in the outer parts of grains such as the bran and the husk. Well-meaning but misguided health food pundits tell you to eat whole grains because they contain fiber and minerals. They don't understand that the phytic acid present in that part of the grain not only prevents you from using the minerals in the grains but also actively removes minerals from your body. Cultures with a long history of eating rice almost universally used white rice when they could afford it; they understood that it was the superior food.

The way we process grains today leaves behind high levels of phytic acid, so we get less of Mother Nature's good stuff and more of the bad. There are more expensive ways of processing grain that remove the phytic acid, but Big Food has largely abandoned those practices. High-speed processing saves money. You are the one who pays the price, because industrial food contains mineral-robbing chemicals that require you to take supplements just to get back to baseline.

My kids have always eaten a high-mineral diet without plant compounds that steal minerals, and the difference is immediately noticeable. When my son was being evaluated before entering preschool, the kindly old teacher picked him up and paused, surprised. "Oh!" she said, "He's heavy, like babies used to be thirty years ago. All the kids today are so light." Kids build strong bones and healthier tissues when they get enough of all of the essential minerals.

Back in the early 2000s, I believed the vegan hype and went on a 100 percent plant-based diet and eventually on a raw food diet. I lasted for about eighteen months. During that time, I had profound joint pain caused by a buildup of oxalic acid in my body, and I started breaking teeth. All of the mineral-robbing compounds in the plants, combined with the lack of animal proteins, depleted my minerals. It took me several years of changing my diet and adding supplements to get to where I am today. By now I've not only restocked my minerals, I've built them up to a truly healthy level.

During a recent surgery to repair an old yoga injury, I heard the sound of the doctor's saw grinding as it struggled to cut through my bone. My surgeon couldn't figure out what was going on. He said to his nurse, "I'm having a hard time getting through the bone here.

What's going on? Is this guy even human?" Afterward, when we talked, he said that he had operated earlier in the day on a patient in his early twenties and that his bone had cut "like butter." That's what eating foods free of mineral robbers can do for you, especially when you supplement the right minerals in the right amounts.

Additional reasons why many people need to supplement their minerals are that many pharmaceuticals prevent mineral absorption and many of us have low stomach acid levels. Part of the way your body processes minerals is by using stomach acid to dissolve them and make them absorbable. As we age, our stomach acid naturally decreases, so the body has a harder time absorbing vitamins and minerals. This is why you might see me pop a few betaine hydrochloride capsules before a meal: they increase my stomach acid to better utilize the food I'm about to eat.

I want you to have all of the minerals your body needs in a form you can absorb. It is one of the easiest things you can do to support an upgrade of your MeatOS. But I recognize that many people find minerals confusing, because there is so much vague (and often overly technical) information out there. In their hubris, health researchers have identified twenty-one "essential" minerals, but that number is somewhat arbitrary. It's not as if you could live without all the other minerals. The main difference is that there is a lot less scientific understanding of the minerals outside the top twenty-one. Let's just call those twenty-one the most important ones, without discarding the rest of the periodic table.

Minerals come in multiple forms, adding to the confusion. You can take them in ionic form (dissolved in water), as a salt (a bonded compound like sodium chloride, or table salt, the chloride salt of sodium), as a chelate (a mineral combined with protein or amino acids), or as a colloidal mineral (tiny bits of minerals so small they don't sink in water even though they aren't dissolved). In addition, there are exotic combinations of minerals with other vitamins that can have profoundly strong effects.

Notably, some of the cheapest forms of minerals are not absorbed well by the body or can even cause harm. Calcium carbonate, a low-cost compound sold as a calcium supplement, is an example; a recent meta-analysis study found that it increased cardiovascular risk for menopausal women by 15 percent.[2] You simply must take minerals together, and you must use the right forms.

One more issue: researchers disagree on how to categorize minerals. Often you will come across the jargony term *macrominerals*. Here is a simpler way to think about the minerals you need and the supplements you should be taking: there are *big minerals* that we need a lot of; there are *trace minerals* that we know we need, but only in modest amounts; and then there are *ultratrace minerals*[3] that the body needs in exceptionally small quantities.

| BIG MINERALS |

There are five big minerals that you simply have to get right, and they come in pairs. *Calcium* and *magnesium* work together. *Sodium* and *potassium* work together. And *phosphorous* works by itself and against calcium. For all of these minerals, it is crucial to consume not only the right quantities but in the right proportions. Too much of one may decrease or increase your need for another.

Calcium

Calcium is famous because it's a major part of your bones, but just as important, it's essential for energy production. Calcium ions flow through cell membranes in order to help generate ATP. You need about 1,000 mg per day, according to the US recommended daily allowance. If you drink mineral water often or eat dairy products (except butter) or drink bone broth, you probably get enough calcium that you don't need a standard supplement.

The most common supplement is calcium carbonate, or oyster shell calcium. That doesn't do very much for you. It isn't worth your time and money. Calcium citrate, formed by combining calcium with citric acid, is another low-cost and widely available calcium supplement. It is absorbed well and will increase your levels. Calcium citrate is useful as a way to get the mineral into your cells; it just provides fewer advantages than the following other four forms of calcium.

Calcium AKG is a powerful way to get calcium into your body, but the calcium is used as a carrier for alpha-ketoglutarate, or AKG, a molecule that helps you build muscle, heal wounds, produce more collagen, and resist the effects of aging. In laboratory studies, calcium AKG increases the life span of roundworms by 50 percent, delays signs of aging, and supports blood vessel elasticity. This is on my essential list of antiaging supplements.[4]

Calcium-D-glucarate is formed by mixing calcium with a compound called glucaric acid. You can get small amounts of this precious detoxing acid from fruit, but you'll get a lot more by taking the calcium salt. Glucaric acid is involved in liver detoxification, so it's important for all of us to have sufficient amounts of it to help us deal with all of the toxins we encounter on a daily basis.

Calcium AEP, or calcium 2-amino ethyl phosphoric acid, is a fantastic form of calcium supplement, because it acts as a cell sealer and protector. Your cell membranes determine what gets through to the inside of the cell. AEP positively alters cell membranes to allow beneficial compounds, such as electrolytes, to enter the cells, while blocking toxins from crossing the cell membranes. Most important, it helps cells maintain their electrical charge by helping calcium and other minerals bind to cell membranes.[5] Because it makes your nerves better at conducting electricity, it has been used with some success in treating diseases such as multiple sclerosis. I take it every day.

Calcium fructoborate is an exceptionally powerful form of calcium that also contains another mineral, boron. Calcium fructoborate relieves the symptoms of physiological stress and is particularly helpful for arthritis and joint degeneration. It also significantly decreases one of the most important lab markers of inflammation, called C-reactive protein, or CRP. It is profoundly more effective than other calcium supplements for promoting bone density.

I don't take "normal" calcium supplements, but I do take these four forms of calcium every day. Choose the one with the most benefits for you, and start taking it. Or take all four, as I do.

Magnesium

Magnesium is a cofactor in more than three hundred enzymatic processes that occur in your body that are responsible for ensuring that you have enough energy. You won't succeed in reprogramming your body's laziness principle unless you have enough of this mineral. Magnesium is also important for making proteins, controlling blood glucose levels, and regulating your blood pressure. It also helps your muscles relax and is very useful if you regularly experience muscle spasms. It's amazing how much a good magnesium supplement can change everything. When I say *everything*, I mean reducing pain, depression, diabetes, and migraines and helping you sleep better. The vast majority of us are low on magnesium unless we supplement.

If you follow a keto or carnivore diet, you may be more likely to be low in magnesium since you'll be avoiding plant-based foods, which are the richest sources of magnesium. Fortunately, if you eat a lot of chocolate, you are likely to have a higher magnesium level. If you drink alcohol, you are likely to have a lower level. You will know when you take too much magnesium because a common side effect is what is affectionately known as "disaster pants." If that happens, spread out your doses over the day.

Any compound of magnesium that ends in the suffix *-ate* will work and will be absorbed well by the body. These include aspartate, glycinate, gluconate, lactate, malate, orotate, and citrate. I prefer to take a mix of all of them because they can operate at different levels of the cells' metabolic pathways. There are a few supplement formulas that mix multiple magnesium compounds, but they are more expensive than single forms. Perhaps the most impactful form of magnesium is magnesium threonate, the form of magnesium that can enter the brain.

Magnesium is a circadian mineral, meaning that you use more of it at certain hours than at others; your body exhibits the highest levels in the middle of the day. I take magnesium in mixed forms in the morning so I have more energy during the day, and I take magnesium threonate at night so that I sleep better.

It is best to take calcium and magnesium at anywhere from a two-to-one ratio to a one-to-one ratio. If you get 1 gram of calcium from all sources in a day, therefore, you would want up to 1 gram of magnesium. Most people can take about 500 mg of magnesium per day without gastrointestinal issues.

TL;DR FOR MAGNESIUM

Take 500–1,000 mg of magnesium in split doses, once in the morning and once in the evening. Aim for a mixture of magnesium forms that end in -ate, such as threonate.

Potassium and Sodium

Sodium gets a bad rap, even though your body needs it to handle stress effectively. The current recommended daily allowance for sodium is so low that it increases heart attack risk by elevating an enzyme called *renin*. The reason salt developed a bad reputation is that it can be harmful if you don't consume the matching amount of potassium to create balance—and many people today are off-kilter on their mix of the two minerals. Sodium and potassium work together to keep your body fluids in balance and ensure that your nerves send impulses correctly.

The sodium-potassium imbalance is a relatively modern development. For thousands of years, salt was a precious commodity. Now that salt is abundant, we eat more of it than we used to. At the same time, we get a lot less potassium in our diet than we used to. This is because our soil today is much less mineral rich and we eat a lot less potassium-containing vegetables than we used to. The imbalance is the reason people become obsessed with cutting sodium out of their diet. Simply raising your potassium intake could solve the problem.

Both sodium and potassium are hydrating minerals, meaning that they draw water. In the morning, I put a pinch of sea salt into mineral water and drink it for this reason. Potassium puts water inside your cells. Sodium puts water outside your cells. When you have too much salt without enough potassium, your cells become dehydrated and you can suffer increases in blood pressure, although very few hypertensive people are salt sensitive; far more are simply potassium deficient. You need salt and potassium together for your neurons to send signals, and when you have enough potassium in your diet, you don't have to worry about how much salt you eat. It's the ratio that matters.

Taking a high dose of potassium is a problem because it can create dangerous fluctuations in your heart rate. That is why most potassium supplements are limited to a tiny, 99 mg size. On the other hand, the government thinks women need about 2,300 mg every day and men need about 3,400 mg per day. Getting that much potassium would require taking thirty-four tiny capsules every day (although in reality you also get some potassium from the plant foods in your diet). Even those numbers are probably too low for most people. A better target is 5,000 or 6,000 mg spread out throughout the day. If you are on a keto diet, you will likely need even more sodium and potassium together.

Do not take potassium if you take prescription blood pressure–lowering medication without talking to your doctor first. If you have kidney problems or regularly take NSAIDs such as aspirin, ibuprofen, or naproxen, ask your doctor first. If you take a potassium supplement and you feel changes in your heartbeat, such as skipped beats or a rapid heartbeat, or if you develop a sense of confusion, weakness, or numbness, stop taking it and talk to your doctor about getting your electrolytes measured.

There are two common potassium supplements that are worth taking: potassium citrate and potassium aspartate. If these two are the only ones you have access to, you're off to an okay start. It's a pain to take fifty tiny little pills to get enough potassium every day, however, so here's a helpful trick: spike your diet with potassium bicarbonate, a chemical cousin of baking soda (sodium bicarbonate).

I take 300 mg of potassium bicarbonate as a powder mixed into water twice a day, away from meals. The reason I do so is that bicarbonate encourages better physical performance[6] and is associated with longer life span. Excessive bicarbonate can cause alkalosis, a dangerous condition in which your body becomes too alkaline. You would need to eat multiple tablespoons of it every day for that to happen, though. There's no need to go crazy on bicarbonate, but it's an easy way to get potassium with extra benefits. To increase your dietary intake of potassium, eat more avocados, sweet potatoes, cooked spinach, and raw grass-fed yogurt if you can tolerate dairy products.

TL;DR FOR POTASSIUM AND SODIUM

Potassium: Aim for a total of 5,000–6,000 mg per day from diet and supplements. Enhance your dietary intake of potassium with potassium bicarbonate powder (I like to take 300 mg twice per day).

Sodium: Don't be afraid of sodium; just be sure to choose sea salt over table salt.

| TRACE MINERALS |

Trace minerals are essential for your body to carry out its energy-producing processes, but you need them in much smaller amounts than the big minerals. That doesn't make them any less essential, though. Trace minerals act mainly as catalysts for enzymes so that you can make energy. Many of them have important antioxidant activity as well. As with the big minerals, both the amounts and the proportions matter, because many of them work together and too much of one of them may increase your need for another.

Copper

Copper is exceptionally important because it helps to protect against allergies and histamine intolerance. People seem to be having a lot more issues with histamine after recovering from COVID; one possible explanation is that to help stave off the disease, many people took excessive zinc, which can create a copper deficiency. If you are low on copper, you might be experiencing a low sex drive, brain fog, less resilience, or excessive urination and more stress. A deficiency in copper also contributes to gray hair and osteoporosis. I recently realized that I was getting excessive zinc because some of the supplements I was taking contained added zinc. After two years, I noticed I had a lot more gray hair. I am currently working to reverse that by increasing my copper levels.

You don't get much copper from food. Unless you eat an ounce of liver, one oyster, or about 40 grams of dark chocolate per day, you are likely not getting enough in your regular diet. The goal is to get 1 to 3 mg of copper per day. Always make sure that you also take an adequate amount of zinc, because copper and zinc balance each other. One of my favorite forms of copper is copper orotate, which is an unusual supplement but is absorbed very well compared to copper glycinate.

However, there is a more powerful form of copper that is associated with antiaging. Scientists have discovered that copper can be bound to a B vitamin called niacin that increases blood flow. The patented combination, called Cunermuspir, from a company called MitoSynergy, is profoundly effective.

Zinc

This is the A-list celebrity of trace minerals. Its fame is well deserved because taking zinc decreases your chances of getting an infection, improves your blood sugar and blood pressure, and slows inflammation. Zinc also keeps your thyroid healthy and helps create sex hormones and adrenal hormones. The problem is that just about every supplement company adds zinc to its multivitamin and mineral formulas, so you may be getting more zinc than you actually need. This can deplete copper from your body, which causes its own set of problems, including gray hair. The goal is to get enough zinc but not too much.

If you eat red meat regularly, you are likely getting enough zinc. The same is true if you eat oysters. However, if you eat processed food or grains, you may be unable to absorb zinc. Do not take your zinc with coffee unless you also take a digestive enzyme containing phytase, because coffee does contain some phytic acid. Ideally, take zinc on an empty stomach or with vinegar or other sour foods that help with absorption. A high-fat keto diet or high-sugar diet requires extra zinc.

Zinc comes in a variety of supplement forms. Zinc gluconate and zinc citrate are most common, but you will see zinc acetate and zinc sulfate as well. Zinc orotate is my favorite form, followed by zinc carnosine, which is specifically to improve digestion. Take two to three times as much zinc as copper, up to about 15 mg per day. Doses above 30 mg per day could lower your blood sugar level or create other problems. It is exceptionally easy to get too much zinc, given how common it is in supplements these days.

Iodine

Getting enough iodine is so important that the US government passed a law that iodine must be added to commercial salt. If we could raise iodine levels globally, it would increase the average IQ by several points around the planet, because it influences brain development. As an adult, if you are low on iodine, you will have low sex hormone levels, hypothyroidism, and thinning hair, and the outer part of your eyebrows may be very thin or fall out. A low iodine level is also associated with depression, frequent colds, joint pain, and tender breasts.

If you do not eat iodized salt, shellfish, or seaweed regularly, you're going to want to take an iodine supplement. If you drink chlorinated tap water or frequently swim in public pools that use chlorine or bromine, you will need extra iodine. If you eat large amounts of cruciferous vegetables, as I did back when I was a vegan, you will need substantially more than the recommended allowance of iodine, which is 150 mcg per day. I prefer to take 500 mcg per day, and some physicians advocate up to 15,000 mcg per day. There is considerable debate about whether people with Hashimoto's disease, an autoimmune disorder that attacks the thyroid, should supplement with iodine or not. Given that iodine plays a role throughout your body, I do not believe that restricting iodine is a great long-term strategy.

You can take seaweed capsules, which contain a predictable amount of iodine, eat seaweed, or take a form of iodine called Lugol's iodine that is sold around the world to sterilize water. A couple of drops in water works well, or you can even paint it on your body, where it will be absorbed through your skin. If you are low on iodine, it will be absorbed quickly. If you are iodine sufficient, you will have a brown stain on your skin for twenty-four hours or more. A more reliable way to test your iodine level is with a urine test.

TL;DR FOR IODINE

150 mcg per day from kelp powder or potassium iodine, more if you are deficient. You can test your level with a urinary iodine test.

Iron

Iron is a double-edged sword. We need it for many essential processes in the body, including energy production in the mitochondrial electron transport chain and carrying oxygen to the tissues. However, because iron can catalyze reactions and create damaging reactive oxygen species, having too much of it in your body can damage your cells, increase your cancer risk,[7] and contribute to accelerated aging. If you have anemia, consider taking copper before you take iron to see if the problem resolves. If you need iron, take an iron supplement—or, better yet, eat more foods containing bioavailable iron, such as grass-fed beef, liver, or oysters. Menstruating women lose blood every month, so they are at a higher risk of developing iron deficiency anemia. For men, a bigger problem is excess iron, which is why donating blood regularly increases a man's life span.

> **TL;DR FOR IRON**
>
> You probably don't need to to take an iron supplement. If you have iron deficiency anemia, consider taking copper first to see if the problem resolves, or consume more whole-food iron sources such as grass-fed beef.

Manganese

Manganese is a trace mineral that helps maintain healthy blood sugar levels, keeps your blood vessels working, and supports healthy joints and bones in addition to basic mitochondrial function. You need about 2 mg per day. You're unlikely to get a consistent amount of manganese from food unless you eat two large servings of plants with every meal.

> **TL;DR FOR MANGANESE**
>
> At least 2 mg per day from food and supplements.

Molybdenum

Molybdenum is important because if you are even slightly deficient it can change your mood, make you feel unmotivated, reduce your stress resilience, and even cause chronic pain and problems sleeping. A lot of common practices increase your need for molybdenum. Hormone replacement therapy or birth control pills, a high-protein carnivore diet, or a vegan diet can deplete your molybdenum. Even so, you need only 50 mcg per day on average, but you can buy supplements of molybdenum glycinate with 500 mcg doses that are affordable. If you are detoxing, molybdenum is even more important because it recycles glutathione, which is your body's primary detoxing substance.

TL;DR FOR MOLYBDENUM

At least 50 mcg per day from molybdenum glycinate supplements and/or food, more if you are on hormone replacement therapy, birth control pills, a high-protein diet, or a vegan diet.

Selenium

Selenium increases your stress resilience and can help your body stave off mercury, an environmental contaminant that is increasingly common. If you get white streaks or spots on your fingernails, you may have a selenium deficiency. I had spots for years, and they went away soon after I started taking selenium. Different parts of the world have varying levels of selenium in the soil. For instance, there is very little selenium in the soil on Vancouver Island, where I live. When a group of Canadian scientists tried to introduce moose to live on the island, the animals failed to thrive because of the lack of selenium. They kept swimming back to the mainland for a better diet. Apparently, they are good swimmers.

Selenium is almost as important as iodine to a healthy thyroid.

And you really want a healthy thyroid, because it is your body's energy thermostat. If you don't make enough heat and energy, you're not going to be upgrading anything. Too much selenium is toxic, with telltale signs being brittle nails and an increase in diabetes and cancer risk. In an ideal world, it's best to get a blood test if you're not sure. Depending on your lab result, a level of about 100 ng/mL is good.

Lots of diet advice sources encourage you to eat one Brazil nut a day for selenium, but that is not something I recommend. If you eat seafood, meat, eggs, and cheese, and sometimes liver, you are probably in the right range for selenium. The best supplemental form of selenium is selenomethionine, taken every other day. Manufacturers sell 50 to 200 mcg sizes.

TL;DR FOR SELENIUM

Get your blood level tested first. If your level is less than 100 ng/mL, take 50–200 mcg of selenomethionine every other day, or get it from dietary sources such as wild-caught seafood, meat, and eggs.

| ULTRATRACE MINERALS |

Ultratrace minerals are essential in your body, but in even lower amounts than the trace minerals. Some of the ultratrace minerals are aluminum, nickel, and vanadium. That list might surprise you, since these are metals that you probably generally try to avoid. It turns out that our bodies actually need them, but the size of the dose determines whether they will be poisonous or not. Like other minerals, they help enzymes do their work to ensure you have enough energy to redirect your laziness principle.

There is an easy way to get all of the vanishingly small amounts of the minerals that your body uses. Around the planet, there are ancient deposits of plant matter from which all of the plant material is gone but the minerals from the plants are left, bound in humic or fulvic complexes. These deposits contain more than fifty different minerals found in plants, which have unique biological properties. They can bind to toxins and excrete them from the body, which is why they are part of what makes my Danger Coffee exceptionally low in toxins. When they are added to coffee, the higher temperature of brewing enables them to bind to toxins, and then the body benefits from getting all of the minerals it requires. You can also buy humic and fulvic minerals as a liquid or in capsules.

Another similar substance is known as shilajit; it likely does not come from plants, but it is secreted by certain rock formations in India and used as a broad-spectrum mineral supplement. I prefer fulvic and humic acids for my ultratrace minerals.

TL;DR FOR MINERALS

Calcium: 1,000 mg per day from calcium fructoborate, calcium AEP, calcium AKG, and/or calcium-D-glucarate.

Magnesium: 500–1,000 mg in split doses, once in the morning and once in the evening. Aim for a mixture of magnesium forms that end in -ate.

Potassium: 5,000–6,000 mg per day from diet and supplements.

Sodium: Don't be afraid of sodium from sea salt.

Copper: 1–3 mg copper orotate per day.

Zinc: 15 mg zinc orotate per day.

Iodine: 150 mcg per day, more if you are deficient.

Manganese: 2 mg per day.

Molybdenum: 50 mcg or more per day.

Selenium: 50–200 mcg of selenomethionine every other day if your blood level is less than 100 ng/mL.

Ultratrace minerals: From Danger Coffee, as liquid, or capsules.

TARGETS AND GOALS

SECTION II

TARGETS AND GOALS

6

PICK YOUR TARGET

Can we be honest? Wanting to be "healthy" doesn't mean much by itself. It's like saying you want to be good. Good at what? One of the reasons your MeatOS hasn't already upgraded you is that it doesn't need to. The best way to survive, on average, is to use as little energy as possible and live long enough to have babies. The bare minimum is enough for the continuation of our species; in fact, sticking to the bare minimum is a damn effective evolutionary strategy. It is just a miserably small, limited way to live. You want more. You deserve more.

Before you can start seriously improving yourself, though, you need to set your targets. As you do so, you will be inundated by messages that will undermine your focus. Some of those distracting messages will come from the outside. You will hear alleged diet experts telling you that being healthy means choking on kale and tofu, while alleged fitness experts tell you that you need to spend huge amounts of time exercising until you want to puke. Other distractions whisper to you from within. Your MeatOS is automatically repulsed by even the thought of any action that requires using more energy than is absolutely necessary. Your body is far more concerned about the F-words. Keeping your head down and not trying anything new, that's Fear. Delicious pizza will make you feel full, that's Food. And sex (or its junk-food equivalent, porn) will feed the third F-word.

All of the steps in this book so far are designed not only to prepare you for hacking your MeatOS but also to get you past those distractions. Once you remove friction, clean up your diet, and add supplements, your body will have a much better supply of raw materials to work with. Your cells will register that you are not in crisis mode. You will have enough resources to give yourself the sharp jolts that will hijack your laziness principle and make it work for you. You will not waste your time and effort by acting on bad advice. Now you can think clearly about what being "healthy" means to you—about how exactly you want to improve yourself.

Picture an energetic guy in a bar throwing darts in all directions. It's possible that one will hit a bull's-eye on the target, but it's a lot more likely that he will spend a lot of energy without much to show for it— other than a lot of pissed-off people around him in the bar. Don't be that guy. Pick your targets carefully. That way, you won't overwhelm your laziness function by trying to do everything at the same time.

First off, take a step back and consider your life goals—not what people have told you to do, not what you think you ought to do, but what is genuinely important to you. It's a lot harder than it sounds. What will you target? What do you want to change or achieve in your life, in your body, in your mind? After a decade of interacting with hundreds of thousands of people about what they really want when they say they want "good health," I have learned that it is always some combination of these five qualities:

- More strength
- Greater cardiovascular fitness
- Improved energy level and metabolism
- Enhanced brain function
- Reduced stress and easier recovery

Often, people also want greater longevity and sexual function, which will happen automatically when you fix these five. Weight loss is another common goal that occurs easily with all five. It turns out that normal people who aren't athletes don't care much about strength and cardiovascular fitness as ends in themselves. What really matters is how much energy you feel or how well your pants fit. A vigorous metabolism conveniently fixes both of those. Sleep is a tool that will help create the five foundations.

Your MeatOS regulates each of the five foundations, and when you improve one, the others will also change over time. All of them will respond to the proper biohacking inputs: a sharp signal to shoot your body up to peak activity and a rapid release to help it come down to a relaxed, powerful baseline. From that list of potential targets, choose where you want to start.

My biohacking path looked like this: Cardio → Strength → Brain → Stress → Energy. Looking back, I regret that sequence, because I was just doing what I had heard was healthy even though it didn't work well for me. If I could go back to my nineteen-year-old self with what I know now, I'd choose Energy → Brain → Strength → Stress → Cardio. Your own hierarchy will probably look different. Choose your goals based on where you are now and what matters most to you!

You're probably all too familiar with New Year's resolution–style promises that you make to yourself about things you want to change. New Year's resolutions set you up for failure because your MeatOS knows that they're going to be hard to stick to, which violates the laziness principle. Well, I've come up with two pieces of biohacking bait your operating system can't resist.

When you choose any of the five foundations, your MeatOS will be inspired by doing things that make you safer and sexier. In other words, you will be addressing Fear and Fertility, which your inner biology salivates for. On top of that, you'll be layering in new ways to improve, so the laziness principle will kick in, too: becoming more efficient will perfectly satisfy the laziness principle's goals. You will have more incentive and less resistance. You can do this!

The power to change is written into your biology. With that in mind, let's take a tour through the main target areas so you can start setting priorities.

CARDIOVASCULAR FITNESS

Some people believe that good cardiovascular fitness is the most important part of their health, usually because they mistakenly believe that it will make them lose fat quickly. Cardio exercise is not a good way to lose fat, but having a good cardiovascular capacity can keep you alive longer. There is evidence that increasing the maximum amount of oxygen your body can process (typically expressed as "VO_2 max") can extend your life span.

Choose cardio exercise as your top goal if you want to spend lots of time running, riding a bike, or hiking. Otherwise, this is good to layer in after strength and stress/recovery. The types of interventions that work well here are HIIT and its newer offshoot, REHIT. With the help of artificial intelligence algorithms, it is possible to improve VO_2 max eight times as quickly as you can by simply doing regular cardio exercise daily.

I've broken down how each hacking target affects the top life improvement goals. These numerical scores will help you select the health target that is most important to you. You can plug them into the Hacks-Goals Matrix on page 126 in this chapter.

On a scale of 1 to 10, improving your cardiovascular capacity impacts the other foundations as follows. Each number indicates the relative intensity of the effect from a cardio boost:

- Energy level and metabolism: 3
- Brain function: 5
- Stress and recovery: 5
- Strength: 3
- Cardiovascular fitness: 10
- Longevity: 8
- Sex: 7

| STRENGTH |

Increasing your strength is a great way to improve your physical appearance, increase your energy output, live longer, and even make your brain work better. It turns out that there is a correlation between the size of your gluteus maximus and the size of your brain. Scientists believe that muscle keeps the brain alive and the brain keeps muscle alive for its own use. It's almost as though your MeatOS wants to make sure that the body it is responsible for has enough muscle to handle whatever is in the world around you. When you have your foundations set up properly, you'll find that putting on additional strength isn't hard because there are all sorts of hacks that work better than picking up rocks and putting them down again, over and over.

Some women are concerned that if they become strong, they will become big. That's not how it works! Additional strength and adequate muscle mass will keep you alive longer and give you a higher quality of life with more energy. Here is how working on strength will impact each of the seven target areas, with 1 indicating the lowest impact and 10 indicating the highest:

- Energy level and metabolism: 7
- Brain function: 7
- Stress and recovery: 4
- Strength: 10
- Cardiovascular fitness: 2
- Longevity: 7
- Sex: 7

ENERGY LEVEL AND METABOLISM

Your metabolism is all about converting air and food into electrical energy for your body. It is the fundamental underpinning of every function of your body. Your metabolism will improve quickly with intermittent fasting, good nutrition, and avoiding toxins that break your cell function. Intermittent hypoxia, breath exercises, exposure to cold and heat, and stress reduction all help as well.

Unless you're already lean and have high energy that doesn't flag, focus on your metabolism early on, because improving your metabolism will cause improvements in almost every major function. In the same style as before, here is how working on metabolism will impact each of the seven target areas:

* Energy level and metabolism: 10
* Brain function: 10
* Stress and recovery: 7
* Strength: 4
* Cardiovascular fitness: 4
* Longevity: 9
* Sex: 7

| BRAIN FUNCTION |

A huge number of people deal with brain fog on a daily basis. I know that I did, and it scared the hell out of me. The good news is that there is a lot of technology that can help you improve your brain at two levels that you probably don't think are possible: you can raise your IQ, and you can fix your memory or eliminate brain fog entirely. If your brain already works well, focus on something else. If Alzheimer's disease runs in your family or you're dropping words or putting your car keys into the fridge, this one is for you.

* Energy level and metabolism: 6
* Brain function: 10
* Stress and recovery: 7
* Strength: 1
* Cardiovascular fitness: 1
* Longevity: 6
* Sex: 7

| RESILIENCE AND RECOVERY |

If you constantly feel anxious and stressed, are having a hard time going to sleep at night, or are just feeling as though you can't handle what life throws your way, you might want to focus on lowering your stress level and improving your resilience first.

Learning to manage stress is important for reaching the ultimate goal of equanimity. When your body is highly stressed, it doesn't take much more stress to throw you into a state of disequilibrium. When you train your body to return to equilibrium more quickly or make it harder for your body to be thrown out of equilibrium in the first place, you're going to feel a new sense of power.

Another common source of stress and anxiety is the brain itself. You can train your brain to be more resilient, and you can train your body to be more resilient. You can even train your nervous system to be more resilient. Keep in mind that almost any of the biohacks in this book introduce stress, but they do it in such a way that your body can return quickly to equilibrium, so it will improve rapidly. That is good stress. Bad stress is when your body never returns to equilibrium or does so very slowly.

- Energy level and metabolism: 6
- Brain function: 6
- Stress and recovery: 10
- Strength: 2
- Cardiovascular fitness: 2
- Longevity: 6
- Sex: 8

| THE CYCLE OF SELF-IMPROVEMENT |

As you read this book, you will find one idea introduced after another, because that's how books work. But if I could seamlessly stitch one end of the book to the other, I could create an even better resource, because that's how your MeatOS upgrade process works. Be prepared so that you can push your body to improve; push your body so that you can recover; recover so that you can begin the whole process again. Respect the cycle, respect the workings of your laziness principle, and you will be amazed at how quickly you can advance toward your targets without feeling resistance or using much willpower at all. Use laziness instead; it's easier!

The MeatOS upgrade is also an interconnected process. Let's say that one day, you're looking to sleep well. That day, you lift really heavy weights. When you do that, your deep sleep is going to increase by thirty to sixty minutes that night, because the exercise will cause your body to need a longer recovery period. Does the exercise cause the deep sleep? Or were you able to do the extra exercise at that level of intensity only because you are already getting enough sleep? There isn't always a clear cause and effect. It's more like herding cats: one does something, then another does something, then the whole group is off in a new direction.

There are many styles of recovery, many varieties of exercise, different times to do things, different ways to eat, and so on. All of the complexity can seem overwhelming, which triggers your MeatOS to shy away from it out of the laziness principle. Don't give in to that. This book is going to save you so much time and energy that even your lazy MeatOS is going to be happy you read it.

In reality, your body's complexity is a feature, not a bug. It means that there is tremendous room for finding the hacks that will work for you. The cyclical nature of the process means that each change will help you make other changes and you will have many opportunities to build on your successes. I'm laying out a wide range of options, depending on your goals and targets. You're not going to do all of them. No one is, *and that is fine*. Just do the ones that will get you where you want to be the fastest.

What if you had the chance to say, for instance: I had one hour a week to exercise, and by working with my operating system instead of against it, I got far more benefits from that hour than I would have before. That alone would be a huge win. Maybe it's because you had mineral sufficiency for the first time in your life so you could make enzymes and build muscles and testosterone. Maybe it's because you exercised in a way that pushed you to your peak output and returned you to your baseline more precisely than you ever did before. Either way, you created more energy with less effort. You changed the one thing that lets you change many more. Then if you change more aspects, you will see more benefits.

If you do even more, you're going to get an even greater payoff. But don't fall into the trap of perfectionism. Don't fall into the trap of feeling that you have to go out and do all the crazy stuff that the billionaires do. My job is to make all the upgrade hacks accessible. On the basis of that, you can set your goals and do what you want to do.

Once you pick your target, you are ready to begin the grand process of fixing and upgrading your MeatOS. The upgrade is a cycle, but it's a one that runs in order. You need to start by removing obstacles and taking in the right resources, or you won't be able to get better at making energy. You need to become more efficient at making energy so that you can have more power. You need to be able to switch that power on and off so that you can recover and reset.

The Hacks-Goals Matrix: Life Goals Versus Upgrade Hacks

At Upgrade Labs, we have an artificial intelligence (AI) system designed to help people identify their end goals and understand their current states. But you don't need AI to get started. The chart below is all you need to begin zeroing in on your own upgrade.

You may have several areas you want to attack. You might want to reduce your stress and anxiety, have more energy, have a healthier metabolism, *and* be strong, smart, and sexy all at the same time. The bottom line is, not everything can be your number one priority. What comes first? You get to decide. You can do it right here in the book if you like.

After years of wanting desperately to improve in every area and being fat, tired, and brain fogged, I can tell you that the universal number one goal for most people is getting their energy back. No one else will know when the energy changes inside you, but you will. You'll find that it's worth it when you wake up in the morning thinking "Today's going to be a good day. Nothing in my body hurts, and my brain's working, I'm not stumbling around, I don't feel hungover, and I don't feel that I have to have a cup of coffee. I'm going to drink one, but I don't have to have it to survive." It's powerful to get to that state where you automatically know "I've got this. I have enough energy to handle whatever comes my way." To do this, start with your cells. If your cells are working efficiently, it will lead to energy.

Here's how to use the matrix. If my number one priority is energy, I'm going to go straight to the "Energy Level and Metabolism" column and look at how all of the hacks rank. Fat-soluble vitamins have the highest impact, so this is where I'll start. I might go back and review the chapter on supplements and start taking DAKE and minerals on a daily basis.

HACKS AND THEIR RELATIVE IMPACT ON YOUR GOAL
(10 IS HIGH IMPACT, 1 IS LOW IMPACT)

Hack	Strength	Energ Level and Metabolism	Brain Fucntion	Stress and Recovery	Cardiovascular Fitness	Sex	Longevity
Fat-soluble vitamins	2	7	9	5	5	7	8
Minerals	6	7	6	7	5	7	8
Brain training	2	5	10	6	3	5	7
Cold and heat therapy	2	4	6	7	5	7	8
Breathing/ hypoxia	1	7	5	4	5	8	7
Light/sound therapy	1	5	7	7	2	8	5

Once you start working on your goal, it is important to measure and track your progress. The easy way to measure your energy is by asking "On a scale of one to ten, how did I feel when I woke up this morning?" You can jot the number down. Then, as you start taking action to upgrade your health at the cellular scale, you can observe the changes in your energy levels and compare them over time.

If greater energy is at the top of your list, what's the second most important goal? Just going through the process of ranking what you want to do will determine your path. It's somewhat of a "choose your own adventure" experience. Once you have your top priority dialed in, go on to your second priority. Remember, upgrading your MeatOS is not a one-size-fits-all journey. Have fun with it.

You are probably doing hundreds of things every day that you believe keep you fit and healthy, even though they don't work very well. That's the human condition. We instinctively believe that progress is possible, yet most of us do not have access to the knowledge to make it happen. Faith in progress is a wonderful thing. Without it, we wouldn't have dishwashers and washing machines. We'd probably all still be rubbing our clothes against washboards in sinks. But faith alone is not much to go on. If you're going to pick your targets and commit to the work needed to hack your MeatOS, you don't want faith. You want some hard evidence that progress is something you can really achieve.

I have spent twenty-five years gathering evidence. I've consulted with leading experts. I've combed through thousands of studies.[1] I've experimented on myself. I created the Bulletproof Diet, which has been used successfully by people all around the world to lose more than a million pounds. I also founded Upgrade Labs and 40 Years of Zen. These companies have put my ideas about the MeatOS and the laziness principle to the test and proven that they work. The companies themselves wouldn't exist if biohacking didn't work well enough to fix my biology and my brain so I could create them. The reason I have my energy and drive is that I've been able to overcome a set of health obstacles that were stacked against me, and I've thrived.

My story is not your story, and my upgrade is not your upgrade.

We are all different. That is why picking the right targets and personalizing your approach is so important. What we all share is the same biological operating system, which has the same quirks, strengths, and weaknesses. You're going to find that the MeatOS hacks outlined in the rest of this book can unleash a level of "normality" that you didn't know existed inside you—if you do them the right way.

Every biohack in here will work better if you consume enough fat-soluble vitamins and minerals because those are the fundamental building blocks of any positive change. We all start out from the same place, the same first principles. You need those basic resources to make enzymes, build proteins, and generate enough energy to get your upgrade going. From there, I want you to be directed in how you upgrade yourself. I also want you to remember *why* you want to upgrade yourself. The ultimate goal is to do what you're here to do, not to spend your whole life upgrading yourself. You will upgrade yourself as you grow and evolve. When you have more energy, everything will work better: your intellect, your moral impulses, even your ability to meditate and center yourself.

Your path will be unique, because every human operating system has its own peculiarities and every person's life is different. If you're a woman in perimenopause looking to lose weight, your path is going to be very different than it was for me when I was nineteen and desperately trying to lose weight. If you are an experienced meditator seeking greater enlightenment, your path will not look much like that of a student trying to keep a clear head during exams.

I can't tell you what your exact path should be, but I can give you the tools and techniques you will need. I can help you choose the path that will be best for you. Your journey begins right here—right now.

7

HACK TARGET:
STRENGTH AND CARDIOVASCULAR FITNESS

With your resources in place and your target locked in, you are ready to begin upgrading yourself. Most people think that exercise is the key to becoming healthier. And for most people, exercise means: picking up stuff, running away from stuff, and that's about it. If you want to become stronger or faster, just do more of it. Bonus points if you drink some bad-tasting protein powder afterward.

But you're smarter than that. You've prepared your body to respond better to signals that give you control of your strength and cardio performance. You've swept away the chemical obstacles that get in the way of your energy. Your body is armed with the right foods, supplemented with the right fat-soluble vitamins, and stocked with enough of the critical minerals to get your MeatOS running at peak performance. You know that there are five primary areas you can target for improvement: strength, cardiovascular fitness, energy level and metabolism, brain function, and stress and recovery. When you send the right signal for your body to improve on one of these areas, the others will benefit, too.

You might be tempted to fall for the work-hard philosophy and say you are in fighting shape, prepared to go to war with the laziness principle built into your body—but you are smarter than that, too. One of the most important principles of biohacking is that fighting against your MeatOS doesn't work. Your operating system is lazy. How long do you think it's going to let you use the energy it gives you to fight against it? It will just make you feel tired, distracted, or bored, and you won't win. You need to harness your biological laziness, not fight it.

Trying to master your laziness by going to war with it would be like trying to keep your Tesla battery charged by driving faster and farther. It's not going to work; in fact, it is counterproductive. The amazing thing is, a lot of people genuinely think it will work. If they are targeting their strength and cardio fitness, they think: Laziness is bad, so obsessive workouts must be good. And workouts are good, so bigger and harder workouts—longer, sweatier, more miserable—must be even better. If a marathon is good, an ultramarathon is better. Struggle causes growth. They really do get sucked into the work-harder mindset.

That mentality is the product of a gaping flaw built in our human decision-making process: we love to be polarized. We sort things into good and bad, and then we get stubbornly stuck on this idea that if something's good, more of it must be better and the opposite of it must be bad. Look back to the original military training in ancient Greece. What did the soldiers do? They trained on one task over and over, doing it every day. It actually works! All you have to do is conscript a huge number of people and have them train all day, every day in military camps, and they'll get strong. Since people did get stronger under that regime, we're still stuck in that style of thinking. Behavior experts tell you that if you want to become good at a task, you should repeat it over and over for ten thousand hours. If you don't do that, it must mean you're lazy and not committed to success.

All of which is nonsense. In almost every system and situation, there is an inverted-U response curve. At first, more input usually does, in fact, produce more results per minute. But then the curve flattens out and drops back down. You reach the point of diminishing returns, when more input produces fewer results. You've overdone it, fooled by the more-is-better ethos. Take drugs and nutrients, for example: a tiny dose of copper or zinc isn't enough to promote health; a medium dose is great; a huge dose is terrible. The same goes for eating, drinking, even breathing.

Exercise is like that, too. Exercise usually follows the inverted U, but AI algorithms and exercise science are discovering all kinds of new ways to get better exercise signals into your cells. What we know for sure is that more is not better. Think about that poor guy who ran the first marathon, or Jim Fixx, who helped launch the jogging craze in the 1970s and then dropped dead of a heart attack at age fifty-two. Better yet, don't think about other people; just think about yourself and what you want your improvement process to look like.

Do you really spend an hour a day in the gym? I doubt it. Even if you do, do you enjoy it? Maybe. But it puts on less muscle per minute than sending the right signal to your MeatOS to transform.

SUPPLEMENTS TO PREPARE YOU FOR YOUR WORKOUT

Coffee

Minerals

Electrolytes or Himalayan salt and lemon juice in water

Essential amino acids

| OUT WITH THE OLD |

My experiments with exercise hacking began many years ago, when I swore I would get stronger legs and lose weight so I could avoid needing yet another knee surgery. I knew that workout time and intensity and starving myself were what would work (after all, that was what it said in a health magazine), so I did all three. I pushed *hard*, with forty-five minutes of weights and forty-five minutes of cardio, six days a week. I did it without fail for eighteen months, and I ate a low-fat, limited-calorie diet. At the end of all that, I was still fat, but I was strong under my forty-six-inch waist.

I decided to take a risk and tried a nongym sport that required twisting. One evening of laser tag took me out of the gym and put me back into line for another knee surgery. After all that work, I still couldn't handle the kind of real-world performance that mattered to me. Now I had to dial back on my exercise routine, too, because my knee didn't allow me to keep up my routine. To my amazement, I started to notice that I had more energy when I was exercising less. Who would have thought?

By the time my knee recovered, I was building my career and commuting to work. I was so busy that I went to the gym only twice a week at most—and I still didn't notice that I was missing anything. That was when I began to realize that I had wasted huge amounts of time and effort working out. I did a calculation: I had wasted 702 hours during that exercise-obsessed period of my life. I still want that time back; it's one reason I started Upgrade Labs. If I had known how to tell my body what to do most effectively, I could have spent the other 500 hours doing something useful. That kind of saving gets my laziness engine all excited!

In this chapter, we'll focus on ways to improve either your strength or your cardiorespiratory capacity, but do it in less time and with less effort. The basic principles you learn here will apply to the other life targets discussed in the following chapters. The old way of strength training was to pick up heavy stuff, such as barbells, rocks, or goats, and do a good number of reps. You might also do a lot of rapid push-ups, some bench presses, or whatever other exercises the equipment would let you do. A trainer in the gym, if you could afford one, might tell you to do thirty reps, forty reps. Do a set of ten, wait, do another set of ten, wait, and repeat.

If you are really committed to it and don't have anything else to do with your time, that kind of training works. You might have to do it every morning at 5:00 a.m., because that's what good people are supposed to do. We do it because it's what body builders in the 1970s figured out worked for them. We've learned a lot since then, but it's probably not what you're doing. In fact, you don't even need weights for the newest, fastest hacks.

Traditional cardio training also causes nasty wear and tear on joints and ligaments, as I learned the hard way. Researchers at Yale Medicine found that at least 50 percent of runners are injured each year, and the number may be much higher.[1] Those injuries can affect you for a long time. You can suffer chronic pain, stiffness, stress fractures, plantar fasciitis, and Achilles tendonitis. In fact, if you run for five years of your life and you have a 50 percent chance of being injured each year, you can pretty much bet on sustaining an injury at some time or another.

It gets worse, because of all those people who believe that if running is good, *more* must be even better! Some of them are harming their hearts with endurance training and marathons, which sucks. Worse still, repeated, long-duration cardiovascular exercise teaches the heart to beat faster and move less blood per beat. The healthiest and most resilient animals (and humans) have a very large ejection fraction. That means their hearts can move an enormous amount of blood with just one beat.

We're stuck in a conundrum: we don't want to beat ourselves up, but we do want good VO_2 max (the ability to consume oxygen) because it keeps us young, and there is even some scant evidence that extreme endurance training may lengthen telomeres, the protective caps at the ends of your chromosomes that protect your DNA from aging. You're going to learn a better way to get the benefits of exercise without wasting time or hurting your joints.

The overall message is: stop fighting your body's laziness, because it has driven biology on this planet for more than a billion years. It will always win over time. Choose what you want most, and let your laziness motivate you. You can get tons of results from very short exercise sessions that don't hurt and don't suck.

| STRENGTH TRAINING HACKS |

BIOHACK HIERARCHY

- **Barbells** used with slow, eccentric movements are easily accessible, but they require cautious use.

- **Nautilus-style machines and weights with cables** are easily available and slightly better than free weights.

- **Isometrics** are great for when you have limited time and equipment and can help you learn proper form, but won't lead to substantial strength gains.

- **Resistance bands** are cheap and easy but are not adjustable.

- **Electrical stimulation (EMS)** produces targeted results, although it is uncomfortable and requires a lot of special expertise.

- **An AI-controlled machine** uses a normal-looking piece of gear guided by a computer algorithm, so it's not moving like anything else does. It gets you the greatest returns with the smallest investment, fewest injuries, and least friction.

Your body has a built-in system of sensors that detect movement, action, and location. Those sensors are called *proprioceptors*, and they let you magically close your eyes and touch your nose. Unfortunately, they prevent you from performing to your full strength. They work with your brain to set fake limits on how hard you can push or pull or lift, somewhat the way your body's laziness system fools you into thinking you have less energy than you really do—except that proprioception pays more attention to fear than to laziness, because its job is to keep you safe.

You can think of the proprioceptors as a highly intelligent system of little nodes keeping track of position and movement. The proprioceptors enable each part of your body to have an awareness of where it is in space. Each individual nerve in a ligament, tendon, or muscle is small and stupid, but the information becomes powerful when your brain uses it. In this way, your ankle thinks for itself, your wrist thinks for itself, and so on. By keeping track of themselves, all the parts of your body can optimize their movements without your being consciously involved. It's another part of your MeatOS that operates completely outside your conscious awareness, like breathing. You don't have to pay attention to it, but you can change it consciously.

Most of the time, the proprioception system takes good care of you. You wouldn't want to have to think about where your ankle is and what it is doing every time you take a step! The system also acts as a safety net to keep you from overexerting and hurting yourself. For instance, your shoulder has a local system that says, "Don't overload me. If I think I'm going to get injured, I'm going to give you pain so that you'll stop." When the proprioceptors send a fear message like that, you are at their mercy. If you try to push your shoulder too hard, it will tell you that you can't do it, and you will believe it. Proprioceptors set those grunting "ugh" limits at the gym, when you think there is no way you can muster any more strength. Training is all about using willpower to push your body a little beyond what your proprioception thinks is possible so that a muscle will adapt (by getting bigger), a joint will adapt (by getting stronger), and your nervous system's sense of proprioception will adapt (by believing that it is safe for the body to lift more).

The thing is, a lot of the time your proprioceptors lie to you. You actually have more strength, more untapped potential, but the conservative limits in your body won't let you tap into it. As soon as you show the body that you can safely get past that level, though, everything changes. The body says, "Oh! I didn't get injured after all." Then it adapts. It resets the limits of exertion. Your body starts activating systems to strengthen the ligaments, which connect bone to bone, and the tendons, which connect bone to muscle, so that you can handle greater loads than before.

Your challenge, as a biohacker, is to find a way to tell your muscles to grow without your proprioceptors stepping in and blocking your efforts—and to do it without harming yourself. Sometimes the body is right, after all. If you are throwing around weights in the gym, you really can cause yourself serious injuries. No wonder your proprioceptors freak out. The reason they're afraid is that they are trained to respond to the acceleration of gravity: 9.8 meters per second squared. Your body may weigh 150 pounds when you stand on a scale, but if you jump off a roof, gravity means your body weighs a lot more than 150 pounds when you land. That's why you won't jump off a roof without a lot of encouragement.

Let's say you're picking up a twenty-pound dumbbell and you wobble it just a little bit. While the dumbbell is still, its weight is 9.8 meters per second squared. As soon as it wobbles, though, there's an extra acceleration on the it and it no longer weighs twenty pounds; it weighs more, just like your body when you hit the ground. The proprioceptors in your wrist, shoulder, and elbow will note the presence of acceleration and automatically limit how much you can pick up, just in case you wiggle and make it weigh more.

So there, dear reader, is our first big hack: Remove or change the acceleration that your proprioceptors sense, so you can push your muscles to new levels in less time without being blocked by your proprioception.

Removing or modifying the force of gravity is not as hard as you might think, because you only have to trick your proprioceptors. Changing gravity itself is a superpower that's beyond me for now, so I'm focusing on what I can do. We know that fully exhausting our muscles in the smallest amount of time leads to the fastest gains. You simply can't fully load muscles that are worrying about gravity.

Starter Hack: Weights, Cables, and Nautilus-Style Machines

These types of exercise equipment work by moving weights against gravity, so they won't trick your proprioceptors. Still, you can get better results by focusing on rapid exhaustion with good form and heavy lifts. You want to take at least ten seconds to lower the weight, also known as the eccentric portion of the movement. Do the number of reps it takes to reach exhaustion (ideally about ten), with no breaks. This still creates a much more time-effective workout than a traditional gym routine does.

Starter Hack: Isometrics

The most famous isometric exercise is the plank, where you pretend to be holding a push-up. Isometric exercises are unusual because your joints don't move and your muscles don't change length. The benefit of this is that you can recruit many muscle fibers at once, and they will help you learn good form. You can do them if you are injured, and they can even help lower your blood pressure.[2] You don't need any equipment. Choose isometric poses without body weight, which introduces gravity. If you stand in a narrow hallway and push as hard as you can against one wall while bracing against the other, there is no gravity involved and you can recruit lots of muscles quickly.

Isometrics don't lead to rapid gains, but they are surprisingly effective in short bursts when you have minimal time and no equipment.

Midlevel Hack: Resistance Bands

The most accessible and sensible way to remove gravity from your workout to get more results in very small amounts of time is to use resistance bands. These cost anywhere from $20 in their cheapest form up to $500 or more for high-quality, high-load bands with special grips. They may seem old-fashioned, but they are quite effective because your muscles are fighting against the elastic resistance of the band instead of against gravity and the erratic motions associated with any kind of free weights.

Gravity is constant, but cheaper resistance bands increase resistance the more they're stretched, which does load your muscles in a way that doesn't resemble gravity at all. The limitation is that stretching the band gets harder the more it is stretched, which means uneven stimulation. It doesn't take much time to exhaust your muscles, which causes faster results. Your proprioceptors start to tell your body that it has to re-create itself to thrive in an environment in which you need extreme amounts of force to overcome that increasing resistance. The end result is muscles that will grow about three times as quickly as they will from lifting weights.

If you want to get more sophisticated, you can upgrade to variable-resistance bands. These are made of layered materials that give you the right amount of resistance you need to lift heavy and to build muscle, whether you are training over a small range of motion or a large one. A while back I had an inventor and trainer named Dr. John Jaquish on my podcast,[3] and he described his version of variable-resistance bands. These bands are designed to give you a lighter load in your weak ranges, normal heaviness in your mid-strength ranges, and a very high load in your strong range. By giving you the appropriate level of resistance throughout your range of motion, you create the perfect environment for building muscle. A bonus is that you can get results in just ten minutes a day.

You can buy versions of resistance bands from low-cost stand-alone bands to complete integrated home gym systems.

Wild Hack: Electrical Stimulation (EMS)

Another, more cheeky way to fool your proprioception is electrical stimulation, or EMS. This technique was pioneered by East Germans and Russians in the 1980s. Russia and the old East Germany were pioneers in biohacking, and Russian science on human physiology and biology is still some of the world's most advanced. The old Soviet Union pushed this type of research because it wanted to win the Olympics. It also wanted to create supersoldiers and superastronauts. Its scientists learned a lot about human performance along the way.

To help athletes train harder and faster, East German medical researchers in the 1980s would anesthetize them and then run large amounts of electrical current over their muscles while they were passed out. It had to be done that way, because the pain was too intense to bear if the athletes had been awake. Even though they were not conscious, their bodies were paying attention and recording that they could handle that level of stimulus. When the athletes woke up and recovered from the anesthesia, they had a greater power output. They could go back to the gym and demonstrate greater strength because they had shown the body it wouldn't break if it got past the proprioception point.

To be clear, I do *not* recommend knocking yourself out and plugging into electrical current the way those East German athletes did. That's because general anesthesia is really bad for you and because you don't need to do it anymore. In 1991, a now-defunct company called Therastim introduced the first mixed-waveform device that had alternating current and direct current at the same time, no anesthesia required. Even though the Therastim technology is off patent, the only available form of it when I started using it was absurdly expensive, so I found a prototype unit from Russia that contained the mixed waveforms. I took it everywhere for a couple years because it was so effective at building muscle and repairing tissue.

Two quick stories to illustrate exactly how powerful electrical stimulation is compared to lifting weights or doing push-ups. One is about my friend the author and performance expert Steven Kotler. Sometime around 2012, when I was carrying my Russian device around, he told me that his shoulder had been hurting for months and nothing had helped. We went up to my hotel room, and I hooked the electrodes to the correct positions on his shoulder. I turned the current up and asked him to move his shoulder. He looked at me and said, "I can't." That was his proprioceptors telling him that if he moved his shoulder, it might get hurt again. They were afraid. So I did what any good biohacker would do: I trash-talked him. I asked him if he was in charge of his shoulder or whether it was in charge of him and probably called him a few names I'm not going to reprint here. He started sweating profusely as his body resisted. Then, with a mighty scream that resulted in some calls to the front desk, he used his willpower to overcome the resistance from his proprioceptors. The second he overcame the resistance, it stopped hurting. I saw him the next morning in front of the hotel, smoking his characteristic cigarette. He smiled, moved his arm freely, and said that it had been cured by my infernal machine.

I also lugged the stim machine to the tenth Ansari XPRIZE event, put on by another friend, Peter Diamandis. Peter created the XPRIZE, which led to private space travel, something that historians a hundred years from now will mark as a foundational change in society. As you might imagine, he had some really cool friends at the party, including a few exceptionally successful entrepreneurs. I hooked my Russian prototype up to their biceps and trash-talked them until they overcame the electrical resistance. Same as with Steven Kotler: the second they realized that their biceps could overcome the current, their proprioceptors stopped complaining and their muscles got stronger and bigger. I became friends with the entrepreneur Naveen Jain at that event and eventually became an adviser to his company Viome, all thanks to a sketchy EMS machine.

Today, there are several companies that make mixed-waveform EMS machines and even some that are attempting to build fitness studios based on the technology (if you want to work out in wet clothes wearing an electrical vest). The most effective and versatile machine I have found after testing dozens is the NeuFit. I recommend going to a trainer or physical therapist who has one of the NeuFit EMS devices, because you will transform your biology much more quickly than you would believe.

Advanced Hack: An AI-Controlled Machine

At Upgrade Labs, we are developing smarter, high-tech ways to get around proprioception using artificial intelligence (AI). The goal is to control the force acting against your muscles in precise ways, tailored to create the maximum response in your body's operating system, so that you will produce the greatest strength improvement with the least possible effort. Weights can wobble and produce strain. Resistance bands, no matter how well designed, can provide only a limited range of forces. With an AI-controlled electrical exercise system controlling the force against you, you can train against what your muscles are doing from moment to moment. You can also watch how you're doing on a monitoring screen, which gives you added cognitive motivation: "Ooh, I'm going to do it. Look, now I'm doing it just a little better than before!"

The Upgrade Labs exercise machines use motorized resistance, not free weights, so nothing ever falls back at the speed of gravity. There is no danger signal to set off your proprioceptors. The machines enable you to put on muscle more than three times as quickly as you can by fighting against gravity, and they are far less likely to lead to injuries.

I had fun demonstrating the technology to Mark Bell, one of the top five power lifters in the world. I put Mark onto a mechanical AI exercise device that we call the "cheat machine." He was pushing numbers I'd never seen before, but the AI algorithm will always win. It doesn't matter who you are or how strong you are, the machine is stronger. It can pick up a truck. After five reps, Mark was blown out and had a look of disbelief on his face. The whole time, I knew he was safe. If he had bench-pressed as much as he was pushing on that machine, he could have wobbled, blown a shoulder, and dropped the bar. It could potentially have killed him, even if he'd had a spotter on each side. With the cheat machine, he got past his proprioceptors. Similarly, I belt-squatted 1,600 foot-pounds on the cheat machine. It permanently flattened out the insoles in my hiking boots. I could do that only because gravity was out of the picture.

Note that the purpose of strength training, regardless of technique, is to build strength. It increases your metabolic fitness and gives you back the bone density you had when you were younger. It gives you the muscles and the ligaments that you need to be a fully functional, high-performing human. If you want to sculpt your body and have ripped triceps, by all means, do so. Go to the gym, become a fitness competitor, get totally swole, do whatever makes you happy. But to become strong and capable, to master your MeatOS, you don't need to be sweating and swearing in the gym. If you focus on creating strong quads, glutes, pecs, and lats, you'll have more than enough muscle. You don't need to waste hundreds of hours a year on exercise.

| CARDIO TRAINING HACKS |

BIOHACK HIERARCHY

- **Steady-state cardio** is stupid and probably counterproductive at high levels. You can do better.

- **Varying-intensity interval training** applies some spikes, but not efficiently, by biking or running steadily, mixed in with high- and mid-intensity intervals.

- **High-intensity interval training (HIIT)** applies on-off training to activate a slope-of-the-curve response.

- **Reduced exertion high-intensity training (REHIT)** most fully applies slope-of-the-curve biology because you return to baseline fastest. It works best.

- **Hyperoxic cardiovascular training** allows you to push harder and smarter, so you can improve your cardiovascular fitness in less time.

Like so many people, I used to do the 1970s version of cardio because I thought it "burned calories." I would hop onto my Cannondale road bike or S-Works mountain bike and pound out the miles in Albuquerque. Or I would hop onto a treadmill with a weighted backpack, set up a 15-degree angle, and go for it. That was a pretty hefty workout. Unfortunately, my results were also pretty typical: I spent a ton of time exercising, but I didn't lose any meaningful weight. I surely developed some cardiovascular conditioning, but not enough to make my brain work better.

Later, a lot of people started doing zone training. I got drawn in by it, too. The idea is that you design your exercise routine to get your heart rate to a certain number and keep it there for an extended period of time. I remember going to the gym and using the machines that were designed for zone training. You would put your hand on the treadmill or the bike and it would read out your heart rate. You can still find a lot of these machines around. Basing your exercise on your heart rate is a sound idea—unless you pick an arbitrary fixed number and worship it.

Those standard types of cardio exercise—typically thirty minutes of running or cycling in order to break into a sloppy sweat—are certainly better than not exercising, but they have some serious limitations. By elevating your heart rate for a long period, you are telling your body that when it is under stress, your heart should beat faster. The thing is, each pump of your heart can move only a certain amount of blood. When your heart rate is faster, the heart ends up moving less blood with each pump. What you are *really* doing is teaching your heart to pump smaller volumes of blood, making it faster and less efficient. This is not good.

Because you wanted to keep going a full thirty minutes, you also couldn't push yourself to your top, peak performance. By the time you ended your exercise, you were probably already fading. Then you had a drawn-out cool-down and recovery session before your body returned to baseline. Each of those steps is the wrong approach for hacking your laziness system. Remember, the body responds most effectively to slope-of-the-curve inputs: a sharp increase to peak performance, a sharp drop-off, and a rapid return to baseline.

Many people love running, regardless of its very mixed benefits, because it releases endorphins, the body's natural opioids. I have family members who are addicted to exercise. If they can't run for forty-five minutes a day, they can't handle life. They can't stop. Workers in high-stress jobs also like to run as a way to burn off excess adrenaline so they can go to sleep. I did an amazing podcast with Lieutenant Colonel Dave Grossman, who told me about how first responders cope with their stress. After a SWAT team goes into a house raid, they often do intense cardio so that stress hormones don't mess with them all day long. If you need to do old-school cardio to cope with your job, go right ahead. If it is more of a casual lifestyle addiction, you should step back and ask yourself whether you could manage your stress in a more effective way.

For everyone else, the vast majority who do cardio because they think it's good for them, you're wasting your time. Long-duration, medium-high-exertion cardio puts you into direct opposition to the workings of your MeatOS. It really taxes your laziness system until the point it pumps out opiate-class endorphins to make you tolerate it. It also makes the body burn carbs, even though you likely want to burn fat.

One improvement you could make is to spend 45 or more minutes at a time (for a total of 180 minutes per week) in "Zone 2" cardio, or about 70 to 80 percent of your maximum heart rate. This is easy to do because your laziness system won't get triggered. It's a slow enough pace that you can easily talk, not superstressful. This combination of low effort and powerful effects makes it a useful hack; it even makes your body grow new mitochondria and burn fat. I first heard about Zone 2 in 2013 from Phil Maffetone, a sports medicine genius who deserves credit for figuring out this unusual way to burn fat; I became fully sold a little later after I interviewed Mark Sisson from Primal Blueprint for his book about it. Spending three hours a week is too much, but it's better than spending three hours a week with fewer results. For this reason, I'm not including Zone 2 as an official recommendation; I want you to have your time back. But if cardio is your jam, go for it.

For the rest of us with busy lives, what you really want to do is almost the opposite of a regular cardio workout. If you want to build your energy and cardio fitness as effectively as possible, you want to be at full intensity for twenty to thirty seconds. Then you have to return to baseline as fast as possible, which will amplify the effect. There are a lot of effective ways to do that, and none of them requires huffing and puffing by the side of the highway.

Starter Hack: Varying-Intensity Interval Training

A varying-intensity cardio workout combines high-intensity intervals with medium-intensity and very-low-intensity intervals. You don't need any special gym equipment, and you can do it on a bike or even on foot. It is a pleasant way to tour, and it is significantly more effective than a thirty-minute steady workout but less efficient at activating your mitochondrial response than other, more finely tuned types of variable training, because you get less of that wonderful slope-of-the-curve biological effect.

For the varying-intensity workout, you bike or run hard for about one to two minutes; then you switch to a medium intensity for a few minutes, giving about 50 percent effort. Then you switch to a very, very low intensity. If you have a heart monitor on your watch, you can wait until you see your heart rate start returning to normal. Whatever your heart rate was when you started your workout, you'll hold back until it comes back to that baseline rate. You then repeat the sequence of high to mid to low intensity for about fifteen to twenty minutes, or four to five rounds.

Midlevel Hack: High-Intensity Interval Training (HIIT)

An important thing I learned in my first ten years of biohacking was that high-intensity interval training, or HIIT, is much more effective than normal running or cycling. John Gray, the man who wrote *Men Are from Mars, Women Are from Venus*, described the process beautifully. Gray is an eloquent biohacker who has been on my podcast several times; over the years, we've become friends. One time he told me, "You know what works a lot better than jogging? Sprint for a minute, like a tiger is chasing you, and then lie on your back and let your heart rate come all the way back down. Then sprint again." At first I thought it was crazy, but I tried it and it actually works. You can feel yourself improve faster.

What Gray was describing was the exact same thing that I call the slope-of-the-curve effect. It is a way to make your metabolic level spike quickly (the tiger is coming!) and drop rapidly down to baseline (while you lie on your back), so that you trick your body's operating system into giving you more energy. What matters is not the amount of time that you spend running; it's the speed at which you can get your body back to normal. That, not the ability to run a long distance, is true cardiorespiratory fitness. Mark Sisson, a multitime Ironman winner, changed the way he trains after learning about the

HIIT approach. He now spends a large amount of his training time moving at the speed of a fast walk, interrupted by short sprints, and he is getting better results.

HIIT is fundamentally better than running the way we've been teaching ourselves to run, but you have to learn a whole new set of behaviors to make it work. Here's a simple starter approach: Go to a park and walk really slowly, slower than you normally want to walk. Think of it as a walking meditation. You can even close your eyes as you take slow steps and feel the ground under your feet. Then sprint intensely, full throttle, for thirty seconds. Then go back to your very slow walk until your heart rate returns to normal. Once it has returned to normal, take off again for another thirty-second sprint. Keep repeating this for twenty minutes total. You will get much better cardiovascular performance this way than you will by going for a jog.

The walking pauses are great because they can act as moments of meditative reflection. What does the sun feel like on your face? Can you hear the wind? How does the ground feel beneath your feet? Your life is completely still and chill. You could walk really slowly during those return-to-baseline moments, but there is an upgrade for this. John Gray was right: lying on your back during your rest is more effective and satisfying than just walking slowly.

When you're lying down, you can come to equilibrium faster, because your heart doesn't have to beat as hard to get blood from your legs back up to your heart. Add some deep breathing and it is just—ahhhhh.

Midlevel Hack: Reduced Exertion High-Intensity Training (REHIT)

Inspired by HIIT, a group of British researchers set out to see if they could find a way to bring similar benefits to people who wouldn't or couldn't do the level of exertion HIIT requires.[4] The scientists found that their test subjects, who were diabetic, were resistant to exercise, and the reasons they gave boiled down to "I don't have time, and I don't get results." They started searching for a "time-efficient alternative to HIIT," and in 2011 they came up with what they called reduced exertion HIIT, or REHIT (somewhere along the way, one of the *I*'s got lost). REHIT consists of two all-out twenty-second sprints within an approximately ten-minute exercise session. Studies show that it is associated with improved insulin sensitivity and an increase in VO_2 max, the measure of how much oxygen your body can use during exercise and a general indicator of overall fitness.[5]

Over the past decade, lots of other people have investigated REHIT and discovered that it really does provide fast, efficient benefits. I've done two podcast interviews[6] with Lance Dalleck, a researcher at Western Colorado University, whose team has validated the concept.[7] He says that REHIT works because of something called rapid glycogen depletion. Glycogen is a carbohydrate, stored in your muscles and liver, that your body can use for rapid energy release. Research shows that glycogen is stored in human muscle bound to water in a one-to-three ratio.[8] That is, for every gram of glycogen you store, you also hold on to 3 grams of water. This is one of the reasons that carbs make you bloated.

If you deplete glycogen rapidly—which is what happens during the short, intense activity bursts in REHIT exercise—the body releases two important signaling molecules. One is AMP-activated protein kinase (AMPK), and the other is peroxisome proliferator-activated receptor gamma coactivator-1 alpha (PGC-1α). You'll come across these molecules a few times in this book, because they are major players in energy production. Both of them tell your cells to make more mitochondria in your cells. Having more mitochondria means that you have more environmental sensors, more manufacturing plants, and more power plants in your body.

Normally your muscles would slowly pull broken-down fat or sugar from the bloodstream while you're doing steady-state style cardio. But when an activity is extremely intense, such as during REHIT, your muscles have to rapidly pull from your glycogen stores. With this come an increase in PGC-1α and AMPK[9] and an increase in the number of energy powerhouses (mitochondria) you have in your cells.

Technically, you can do REHIT with a normal exercise bike, but it is hard to change most bikes' intensities quickly enough. If your bike can do it, all you have to do is warm up at a very slow pace for two minutes and then turn the resistance up instantly to do an all-out, 100 percent maximum-power sprint for twenty seconds. After twenty seconds, you'll turn down the resistance and go back to an extremely slow cycle for three minutes. After those three minutes, you'll repeat the all-out twenty-second sprint. This is followed by a very slow three-minute cooldown. A big problem with this DIY approach is that you don't know how high to set your maximum resistance. After all, your laziness genes will kick in and convince you to set it too low. Your best bet is to have a computer tailoring the resistance and difficulty to your fitness level. Fortunately, a company called CAROL makes an AI bike for home use that will maximize your slope-of-the-curve response to create rapid glycogen depletion and mitochondrial biogenesis.

You might want to choose to do REHIT over HIIT. Studies show that because regular HIIT requires such long recovery periods, it isn't as time efficient as it is often claimed to be, and having to repeat sprints more than four times can cause people to have a negative view about high-intensity exercise[10] and therefore less likely to stick with a program. REHIT is over so quickly that you don't even have time to think about how intense those twenty-second sprints feel.

Advanced Hack: AI-Guided Cycling

You can go full high-tech biohacker and do a more advanced form of REHIT by adding machine learning and AI. This is a concept we are exploring now at Upgrade Labs, installing custom algorithms on an AI-enabled exercise bike. People come in and do an eight-minute guided REHIT training. There is no sweat involved. You don't have to even go to maximum exertion for a full minute, because it turns out that a minute is too long. All we care about is how quickly you can go to full power—the slope-of-the-curve effect. The important thing is to be moving at an almost imperceptible pace and then—*bam*—hit it hard for twenty seconds. And you do this only twice.

"Hard" means different things for different people, so the Upgrade Labs bike adjusts to your capabilities and your output, all while monitoring your heart rate. Then the power turns off and you're guided by a computer-generated voice. The AI determines how quickly you should return to baseline, based on how hard it is for you to climb the curve of exertion and how hard it is for you to descend. The AI constantly adjusts what it tells you to do. This customized REHIT technique is incorporated into an overall cardio workout that is five to eight minutes long, but very little of that time is spent sprinting. Most of the time you are creeping along, doing another version of the meditative walk.

Advanced Hack: Oxygen Hit

Back when I was a cardio freak and not seeing much in the way of results from all my running and cycling, I was always looking for shortcuts. And honestly, I still am today. After all, biohacking is all about finding shortcuts and a willingness to experiment. So to speed up my cardio workout, sometimes I'm willing to strap on a mask and inhale a dose of pure oxygen.

The purpose of the oxygen mask is to get more energy in less time. I run companies, I write books, I host podcasts, and above all I am a dad. I have life priorities that are a lot more important to me than working out, and I'm sure you do as well. Frankly, hanging out with my kids matters to me a lot more than spending an hour grinding away on a bike. That's where the oxygen mask comes in.

One of the factors that sets the limit for how much energy you can produce is your VO_2 max. Studies show that breathing hyperoxic gas (air that has a high oxygen content) enables higher-intensity training and creates significant improvements in power output compared to breathing regular air.[11]

Wearing a mask connected to an oxygen supply helps you push through an intense workout and get benefits in a shorter amount of time because of the increased oxygen delivery to your tissues. According to scientific studies, increasing your VO_2 max by 12 percent adds two years of healthy life span.[12] If you can do that, you will also lower your risk of developing type 2 diabetes by about 60 percent. I first tried it in in 1995. I bought an oxygen tank on Craigslist, lugged it to the local big-box fitness chain, and hit the treadmill. It was a big tank, about three feet tall, and heavy. People looked at me weirdly because I didn't seem to have emphysema, given that I was at the gym. When I ran out of the oxygen in the tank after three rounds of training, I couldn't refill it without a prescription. Thus the experiment ended, and the empty tank went for sale online.

It's a little easier now. You can upgrade your exercise routine with an oxygen trainer if you have access to an Upgrade Labs or a sports training facility that offers Exercise with Oxygen Training (EWOT) near you, or you can do hyperbaric oxygen therapy, where you get into a pressurized chamber and breathe pure oxygen.[13] Studies show that hyperbaric oxygen therapy can reduce inflammation,[14] increase mitochondrial biogenesis,[15] and improve VO_2 max,[16] which makes it a great add-on to your exercise and recovery routine. There are many facilities to which you can go to receive hyperbaric oxygen treatments, or you can purchase your own chamber for at-home use. The cheaper models often have problems with off-gassing from plastics. I recommend OxyHealth chambers.

| MOVEMENT HACKS |

BIOHACK HIERARCHY

- **Breathing and functional movement exercises** such as practicing "beer can breathing" and doing active straight-leg raises and neck exercises can help you upgrade your movement without having to leave your house.

- **A functional movement expert consultation** involves going to an expert to assess how your body moves and to address and fix any inefficiencies.

Although it's not normally a part of strength or cardio training, fixing the way your body moves is critical for maximizing your workout and carrying the benefits with you at all times.

Unless you are a professional dancer, martial artist, or yoga teacher, you are likely to have movement problems that you are unaware of. When you take a breath, you don't know that you are forgetting to fill up the back part of your lungs, because your body is not used to doing that. When you put your foot down onto the ground, you probably put the wrong part of your foot down, or you don't using your toes for your full power. If you see a functional movement expert, he or she can often add a foot or even eighteen inches to your long jump in one session, just by helping your body remember how to use certain muscles that it has forgotten about.

Some of your movement problems aren't even yours, really, but inherited from your parents. A bit of that is genetic, but it happens mainly because babies watch their parents move and then mimic them. If your dad walked like a monkey, you're going to walk like a monkey. In my case, I was duck-footed as a kid, just like my dad. If someone had sat down with me for three hours when I was twenty and showed me how my foot was supposed to move, it would have saved me huge amounts of wasted exercise effort. I could have activated more muscles in my leg at the same time. I could have done more high-intensity exercise. Or imagine if I'd had a schoolteacher who'd told me, "Relax your shoulders, pull your shoulder blades together and down your back, and sit without sticking your chest bone out. It's good for you."

In reality, none of us was trained in proper body movement. A lot of us get upper back, shoulder, or neck pain because of the way we work at our desk. We never learned how to move in school, so we carried our bad habits with us. I've seen amazing results from a one-hour session with a functional movement expert. One of them was working on my foot and noticed a muscle I was not activating. He tapped on the muscle, and I was amazed that I could then turn it on and move it. Functional movement "downloads new software" to your body's operating system so you can access a muscle that you didn't know was there.

I first experienced this switching-on process when I started doing intense yoga, sometimes four or five times a week. I was doing downward dog, a really straightforward pose, when a yoga instructor came over and pushed on a part of my back. All of a sudden, my pose drastically improved. When I was doing the seated forward fold, which has always been hard for me, I had one teacher walk up, tap my lower back twice, and pull up on my shoulders. I fell forward six inches farther than I'd ever gone. That gave me a whole new level of insight into the ways that the body doesn't know how to talk to itself. You may feel that you're in full control, but no one gave you the software upgrades needed to address every muscle in your body consciously.

As a rule, if you have chronic pain somewhere in your body, it sucks energy from you all the time. That is energy you could be applying to all the other things you want to do. In addition to the energy suck from the pain itself, chronic pain sucks your energy by making you move incorrectly. If you can afford a massage, you can afford a functional movement consultation. It will make all of the other aspects of your personal upgrade easier.

Starter Hack: Breathing and Functional Movement Exercises

You can start improving the way you move on your own using something called the C-clamp exercise. To do this, you stand up and make a C-clamp with your hand, right underneath your rib cage. Your thumbs point toward your spine, your fingers are in the front, and your back is straight. Then you inhale. Another name for this is "beer can breathing"; it was developed in part by a neurosurgeon named Dr. Marcella Madera together with Dr. Joe Dispenza. You breathe in through your nose, and your goal is to inflate the "beer can," the C-clamps you have formed with your hands that are around you, so that your front and your back fill up evenly. Practice this until you are able to become aware of filling your front and back up evenly without having to use your C-clamps.

Most people breathe through their stomach because they've

learned to do abdominal breathing. They look kind of pregnant when they breathe in, they are hunched over, and they are not breathing in through their back. A functional breath, one that increases your oxygen intake, is where you breathe and your back ribs expand as much as your front ribs do. When you breathe that way, you feel profoundly different because you're getting oxygen to the deeper parts of your lungs. Each breath fills you up in a way you haven't felt before. When I started doing this, I realized that I hadn't been filling the lower back of my lungs, ever, because it wasn't on my radar. Now when I take a deep breath, my back ribs move and my breathing is better and deeper than it was. I also find that my posture is much better, which is essential for proper movement. With more air comes more oxygen, more energy, and more life.

Along with paying attention to your breath, you'll want to address the way your body moves and what might be holding you back from getting the most out of your exercise. The following simple diagnostic trick can tell you if you have a problem with functional movement. You stand on a stool and you straighten your legs, not quite locked. Then you bend down and reach down with both hands. Are your arms exactly the same length? If not, something is holding you back. It's either an injury that your body is compensating for, or it's a functional movement problem. A functional movement expert (whom I recommend seeing, as an advanced biohack) could recommend the proper countermovement, and it probably wouldn't take long at all to fix. You should also have an expert check your shoulder mobility. I found that my left shoulder was seriously lacking mobility. I did some exercises leaning up against the wall that fixed the problem. It's one of those hidden glitches in your operating system that you want someone to uncover so you can fix it.

One exercise that is really useful because we spend so much time sitting is called an active leg raise. To do this, you lie flat on your back and raise one leg as high as you can while keeping your back and your butt on the floor. You may find that one or both of your legs are exceptionally weak and you can do it, but someone putting a tiny amount of pressure with their finger on your foot could stop your leg from moving. What's going on is that either you don't know how to turn on your hip flexors all the way or you have problems with your hamstrings. That's a trainable fix, and if you do the active straight leg raise regularly, you'll probably find that it becomes easier and that your hip flexors and hamstrings become stronger and more flexible.

Another common issue that results from our bad habits (in this case, looking at our phones) is limited neck mobility. To train your neck muscles and treat pain, you can wear a head harness with a resistance band attached to it and nod your head up and down. This can strengthen the deep neck flexors that help you maintain proper posture and stretch the muscles in the neck to improve your range of motion. You can find special resistance bands for your neck muscles online.

Advanced Hack: Functional Movement Consultation

Compared to running, biking, or lifting weights, going to see a movement expert may seem exotic, maybe even a bit silly. In reality, it's a very-high-return-on-investment activity. A functional movement assessment usually takes about an hour. An expert will diagnose which muscles you don't know how to activate; it's like a health screening. It's easy to tell if you're limping, but it's not so easy to tell if you're just walking funny or if you have something serious going on that you've never addressed or even looked at.

This is a great way to get rid of pain, but even if you don't have a lot of pain, it's a great way to get rid of inefficiencies. One to eight sessions with a functional movement specialist (you can find one near you with a quick internet search) will unlock abilities you never knew you had. You'll change how you walk, how you stand, how you move your arms. It doesn't have to be for a specific sport. If you're looking to get more exercise in less time, what could give you more exercise than being able to turn on muscles you weren't exercising before?

HACK TARGET:
ENERGY LEVEL AND METABOLISM

If you ask people what they want from their exercise or food choices, the most common answer you'll hear is a variant of "I want to lose weight and grow abs." Sometimes people also want to be stronger or have better cardio endurance. Even then, though, when you dig deep, the goal behind those goals is often "lose weight and grow abs."

If those are your biohacking goals, I encourage you to ask yourself: What would happen instead if my goal was to be better at converting air and food into electricity? That is where your energy actually comes from, and if you become better at that, you'll have more strength. More cardio. And maybe even some abs.

When your goal each morning is to have more energy, that extra energy flows out of your waistline and ends up changing everything else in your life. Every morning, when you wake up, either you have enough energy to decide who you will be and what you will do, or you don't. If you don't, you'll live on autopilot.

If a high energy level and metabolism are your goals, you need to come up with a signal or input that tells the cells of your MeatOS that they are facing some kind of hardship. If your cells think they have to deal with extreme conditions, they will ramp up their mitogenesis, the ability to grow new mitochondria, and mitophagy, the targeted destruction of weak, inefficient mitochondria.

You can undergo the hardship, starving and torturing yourself, fighting against the laziness principle. Or you can use technology, strategic signaling, and slope-of-the-curve biology techniques to make your body think that it's doing extreme things, so that you don't have to suffer as much. I expect that the second one would be your preferred choice. Right? Then let's go.

SUPPLEMENTS TO SUPPORT YOUR CELLULAR HACK

Active PQQ

Acetyl-L-carnitine

Oxaloacetate

Methylated B vitamins

NAD+ precursors

| VIBRATION HACKS |

When I was a child, maybe about eight years old, I was fascinated by the exercise rooms I saw when I stayed in hotels with my parents. A lot of the time they'd have these strange-looking devices where you'd stand with a vibrating waistband around your midsection. You can find old videos online of people from the 1950s going to clubs where they would be vibrated to lose weight. Even at the time, people laughed at the devices. They didn't know that the idea for the machines had originated from the famed inventor Nikola Tesla and his contemporary Royal Rife, both of whom were using vibration to improve people's health.

Around the time the goofy-looking vibration belts came out, researchers who were studying the limits of human performance—mostly in the USSR and East Germany, but also at NASA in the United States—saw a fascinating potential in vibration: they recognized that vibration might provide a way to send a signal into the body that would help it recover more quickly.

Russian scientists were especially intrigued by the idea of using whole-body vibration to restore bone density in astronauts. It turns out that standing on a whole-body vibrator does some very helpful things to the body, especially when it is vibrating thirty times a second, or 30 hertz (Hz). The reason it works is that your bones are not just inert pieces of calcium chalk; they contain a collagen matrix, special cells called *osteoblasts* that build up their structure, and other cells called *osteoclasts* that carve away and dispose of excess bone. Bones are living tissue. They grow and they shed just like everything else in the body, and they also respond to signals.

Most of the cells in your body are piezoelectric,[1] which means that when you move them or vibrate them, they generate a little jolt of electricity, which simulates growth. When you stand on a whole-body vibration platform, you unleash the piezoelectricity in the cells of your bones, promoting growth and wound healing[2] (just as the Russians suspected) while also building bone density. Those old vibration machines weren't so stupid after all!

Whole-body vibration also puts oxygen into parts of the body where there normally wouldn't be much. It accelerates drainage from your lymphatic system (I'll share more about that below), which helps release toxins from your body. Perhaps the most interesting thing about vibration is that it tricks your body into thinking you're doing way more movement than you really are. If you stand on a whole-body vibration platform and do a squat, your brain and your muscles think that you're doing thirty squats a second, because they're constantly adjusting your movements. If you try to hold a yogic plank pose on one of those platforms, you'll get a crazy workout.

BIOHACK HIERARCHY

- **Vocal vibration,** such as chanting or humming, to induce nervous system relaxation.

- **Rebounding** on a minitrampoline can help with circulation.

- **Spot vibration** with vibrating foam rollers and balls can help activate tissues.

- **Whole-body vibration** involves standing on a plate that shakes your whole body at 30 Hz to increase lymphatic system circulation, improve bone density, and enhance your workouts.

Superaccessible Hack: Vocal Vibration

What's that you say? Dave, I'd like to try something *free*? There is one type of vibration therapy that costs nothing at all: good old-fashioned chanting. You sit down, take a deep breath, and say "Ommmm." When you resonate your chest that way, you feel your whole body hum. You might feel like an idiot, but who cares how you look? If you happen to have a baby, hold the baby in front of you while you say "Om," resonating your whole chest cavity that way, and just watch what your baby does. Babies are fascinated by the sound, and they relax completely. I did it with my kids, and it was a wonderful experience.

There are many books and websites that tell you how to chant, so many that you could write a volume of books on it. I'm a fan of the Art of Living, which includes breathing and chanting and is used by tens of millions of people a day, or just going to a yoga class that ends with chanting a few phrases. It works.

Before I had kids, I would wake up at five in the morning and chant "Om" for five minutes or practice other advanced chants by Dharma Singh Khalsa, an American doctor who became an expert in chanting and became so convinced that it worked for medicine that he became a Sikh and changed his name. It made a big difference in my energy. Deep resonant bass at certain frequencies has a profound healing and energizing effect.

Musical sounds can vibrate you in much the same way as chanting does. Some people swear by the didgeridoo, an Australian instrument that makes a deep, booming sound. You can find healers who will play a didgeridoo for you or learn to play one yourself; very basic ones start at less than $50. Other low-cost hacks are to take a "sound bath," use a "singing bowl," or buy a sound device that you can switch on when you go to bed. It'll help you sleep better, and when you wake up, your nervous system will be refreshed. If nothing else, humming a tune helps.☺

Accessible Hack: Rebounding

If you want to hack your cells with vibration—and especially if you want to target your muscle cells more than your nerve cells—the lowest-hanging fruit among commercial vibration devices is a rebounder. You have definitely seen these. There's a decent chance you already have one in your house or apartment or maybe languishing forgotten in your garage. A rebounder is like a little trampoline that you can keep in your living room; they were popular in the 1980s.

Years ago, I was invited to speak at a big public event by Tony Robbins, the life coach and motivational speaker. Before it started, I went backstage, and there was Tony, jumping up and down on a rebounder. This is a simple but effective hack because when all the tissues in your body move and recirculate, your lymphatic system moves and oxygen moves. At the same time, all the muscles in your body and in your fascia get a signal: "I need to be able to grip, release, grip, release." That's another way to trick your laziness system into building up your cellular strength.

If it's good enough for Tony Robbins, it's good enough for me. You can pick up a basic, generic rebounder for about $70.

Midlevel Hack: Spot Vibration

There is a variety of new vibration therapy devices available that use foam rollers and balls that vibrate and perform tissue activation. There are also devices that immerse you in a "sound bath": you lie on a lounger or a massage table that has heavy-duty subwoofers built in, and these vibrate your whole body in an intense way. You can have all sorts of profound experiences doing this. You wear headphones and rest peacefully while your whole body shakes. This is an exciting area of research, and over the next decade we're going to learn much more about how different frequencies—including sounds—affect your tissues as vibrations.

Your cells and nervous system are highly sensitive to vibrations, not just because the motions help deliver nutrients to the cells, but also because vibrations communicate messages about environmental changes. A number of companies are working on small devices that vibrate over acupuncture and acupressure points on the body. The frequency and location of the vibration can have significant effects on your nervous system. For instance, even a small amount of vibration on your wrist can activate oxytocin, the so-called love hormone, that is associated with the feeling of romantic attachment as well as social bonding. Vibration can also shift your body into reset or restore mode, instead of fight-or-flight mode. This deep, profound restful state of healing changes the slope of the curve after stressful exercise. A faster return to baseline equals faster recovery from anything.

Small devices that vibrate specific parts of your body start around $200. Two of my favorites are the Sensate and Apollo devices, which I'll discuss in chapter 9. They function primarily as nerve stimulus devices, but they also provide helpful vibration therapy.

Advanced Hack: Whole-Body Vibration

A whole-body vibration device is a bigger investment, but there are good reasons why you might want one of your own. I had my first direct adult experience with whole-body vibration when I was recovering from toxic mold and a member of the antiaging nonprofit I worked for handed me a key to a small room in a nondescript strip mall in Menlo Park, California, saying, "This will help you." Inside the room were four incredibly expensive whole-body vibration machines, each worth about $20,000. Over the next three months, I got to play with them. Each time, I could use a different frequency and feel what it did to my body. One frequency (below 10 Hz) caused an immediate need to run to the bathroom. Another made me profoundly tired. Others seemed to concentrate in my gut. It was abundantly obvious that different frequencies have different effects.

Years later, after I had created a vibration platform for one of my companies, I was filming my very first training class on the Bulletproof Diet. I had just flown across the country from Florida to San Francisco to record the training video. The photographer and videographers asked me to hold a plank pose on a vibrating platform to demonstrate it as an example of a biohack. I held the pose for about ten minutes as the machine hummed away, far longer than any sane person would have done, because we were trying to get a good shot. Two days later, embarrassingly, I went to the hospital with chest pain. I had so overtrained and stressed the muscles and joints in my sternum that I felt as though someone had kicked me in the chest.

This won't happen to you with normal vibration, but there are two things to watch out for. The first is standing with your legs locked, which can rattle your brain. Don't do that. The second is that some of the cheaper machines rock back and forth from left to right. That is almost guaranteed to wreck your low back over time. Stick with models that vibrate all of you at once, and you'll notice a big difference.

On a macro level, when you stand on a vibrating platform, you are exercising your whole body more quickly. On a micro level, vibrations create neurological changes throughout your system. With some high-end vibration devices, you lie on a bed that transmits vibrations via a powerful subwoofer that that plays specific frequencies to elicit different states of consciousness and types of cellular activity. Some researchers have been experimenting with frequencies that specifically promote bone density,[3] building on the work done by the Russian scientists more than a half century ago.

If you look at a video of cells being cultured in a laboratory, you will understand why vibration has such a great effect. Cells rely on back-and-forth circulation of the liquid they are growing in to keep nutrients running over them. In your body, that motion usually comes from the flow of blood; for single-cell organisms in the wild, it could come from currents in a pond or stream. In fact, when scientists try to culture human cells, they often won't grow unless they're placed in something that moves gently, oscillating them back and forth. Cell movement stimulates cell growth, which in turn creates more mitochondria. A good dose of whole-body vibration mimics the effects of vigorous exercise and heightened blood flow past your cells. When you do whole-body vibration, your whole body feels tighter and leaner because of the lymphatic drainage.

If you are purchasing a whole-body device, you want to get one that moves up and down or one that oscillates in a small circle, not a device that rocks you. Beware of the cheap vibration devices you can buy online, which are usually the rocking kind; they can harm your low back and hips. Even with a high-end device, you need to be careful. You don't want to place your head on it, lock your knees while you're standing on it, or allow it to vibrate your eyeballs. You can cause neurological or vision damage that way. When you stand on a vibration platform, you should always make sure your knees are bent a little bit.

Used correctly, vibration therapy can do a lot to elevate your energy and metabolism. If you normally sit all day, as many of us do in our work, you can reset by standing on a vibration platform for one to five minutes. While you are on it, you can do a couple squats or a couple stretching poses. The vibration will rejuvenate your tissues way more quickly than going for a walk would. You could do ten body-weight squats on the floor, or you could upgrade them by doing them on a whole-body vibration machine. Your squats will be more effective, and you'll get greater tissue circulation and oxygenation.

You can buy machines for home use starting at $500, though most good models are closer to $1,500 or more. Some trainers, gyms, and biohacking upgrade centers have them available for use.

| BREATHING HACKS |

BIOHACK HIERARCHY

- **Breathing exercises,** such as holding your breath, can help you form new blood vessels, and Wim Hof breathing can help your cells become more efficient at pulling in oxygen.

- **Tech-guided breathing** involves changing the percentage of oxygen in the air to reduce your blood oxygen level and increase your cellular resilience.

- **Induced hypoxia** with a hypoxicator, a breathing apparatus that alternates your air intake between oxygen-rich and oxygen-poor air, enhances your body's production of red blood cells.

Accessible Hack: Breathing Exercises

You can send powerful signals to your MeatOS that it needs to change the way it executes its operations simply by controlling the way you breathe. One way to do that is by putting yourself into a state of controlled hypoxia, so that your body doesn't have access to enough oxygen (or at least thinks it doesn't). Extreme hypoxia can create a systemic broad, inflammatory, and harmful response; in its most intense form, of course, hypoxia is deadly. You want to do it in a controlled, brief way to improve the efficiency of your metabolism.[4] This way, you will teach your body to function with low oxygen levels,[5] but only for a very brief period so you do not cause systemic harm. It's using slope-of-the-curve thinking to signal your cells to improve.

Members of cycling teams and long-distance runners have long made use of hypoxia to achieve higher performance. They often train (or live) in a high-altitude location such as Boulder, Colorado, or

Albuquerque, New Mexico, where the air is significantly thinner than it is at sea level. This forces their bodies to adapt to taking in less oxygen. After six weeks or so, their red blood cells start to change and the hemoglobin—the molecule that receives and transports oxygen in the blood—binds more aggressively with oxygen molecules so that it can get more of them to their tissues. Then, when they return to sea level, the air they breathe feels positively rich and overloaded with oxygen. They feel as though they could bike or run forever.

You can summon up that feeling, too, and you can apply it to anything in your life. You don't need to pack up and move to Colorado; you can control your oxygen intake right now, wherever you are. For instance, you could practice Pranayama breathing methods[6] derived from ancient Chinese, Tibetan, or Indian traditions.

Well-designed breath holding can trigger angiogenesis, the formation of new blood vessels. It can also strengthen your mitochondria,[7] increase your hemoglobin production, improve the plasticity of your brain, protect your cardiovascular system, and increase your level of nitric oxide, a signaling molecule that can dilate your capillaries to improve blood flow. Breath-holding exercises are also fundamentally safe (don't do them underwater, though); your physiology makes it impossible for you to hold your breath to the point of unconsciousness. A study out of Duke University indicates that breath control swimming[8] is an effective way to do hypoxic training,[9] which may be one of the reasons that swimming is so powerful at improving your health. Swimming also gives you cold therapy at the same time as you're holding your breath. You just have to be extra-cautious about what you are doing whenever you are in water.

A particularly potent breathing practice involves hyperventilating so that you take in lots of oxygen, exhaling fully so that your lungs are empty, and then doing push-ups or another form of exercise so that your body digs deeply into its oxygen reserves, pushing up the slope of the curve. I once demonstrated this technique onstage with Wim Hof, a renowned extreme athlete. I did push-ups with my lungs empty, no oxygen coming in, and used up all the oxygen in my cells. Then I kept doing push-ups; it may seem impossible, but you can go for a minute and a half doing push-ups with your lungs empty. By then I had fully depleted the oxygen in my cells, sending them an intense signal to stop being lazy and start getting better at pulling in oxygen.

Combining exercise with Wim Hof's[10] breathing technique is another way to induce brief but intense hypoxia at no cost, at any time, in any location—at least, anywhere you feel comfortable doing push-ups with a beet-red face. This enables your cells to become more efficient, increasing your overall energy levels.

Advanced Hack: Tech-Guided Breathing

At Upgrade Labs, we have technology that measures your blood oxygen levels while you breathe from a mask that actively adjusts the mix of gas. We can change the percentage of oxygen in the air to reduce your blood oxygen levels to the point where your cells activate and then bring you right back up. It's a more accurate, responsive way to tap into slope-of-the-curve biology. You may be able to find a biohacking or exercise facility near you that provides this type of training.

If you are doing breath work on your own, you should do it five days a week or more to be effective. If you go the high-tech route, on the other hand, even two days a week is enough to have a significant effect. Once you've done six to eight weeks of breath work, you're probably acclimated, just like those cyclists with their high-altitude training. You will feel a very noticeable change in your energy because hypoxia raises your levels of a molecule called *AMP-activated protein kinase*, or AMPK. That molecule is good to have around because it transports energy-rich glucose into the cells and helps protect your mitochondria from stress.

Advanced Hack: Induced Hypoxia

If you want to go all in with your breathing hacks, you can try out a hypoxicator, a breathing apparatus that alternates you between oxygen-rich and oxygen-poor air. This technology is a modern twist on the old technique of Tibetan fire breathing and the breathing exercises developed by Wim Hof and the health journalist James Nestor. The hypoxicator is based on more early biohacking work by the Russians. In this case, Russian scientists were looking for ways to alter the biology of jet pilots so that they wouldn't need an additional oxygen supply even when they were flying at high altitudes.

Used properly, a hypoxicator forces your cells to work better and your brain to work better. Induced hypoxia elevates levels of a hormone called *erythropoietin*,[11] or EPO, which in turn elevates the production of red blood cells. EPO is the stuff that the cyclist Lance Armstrong used to get higher performance.

In the simplest form of a hypoxicator, you can wear a mask that restricts your breathing when you're exercising. You might see some people running around looking like Bane from the *Batman* movies, using old-style masks from a time in biohacking when it was a bit of a trend. The new masks look less frightening and feel more natural, but they are still quite a bit of work to use. They are also risky if not used properly. Viome, a company that I'm an adviser to, did a study and found that long-duration hypoxia (in which you spend a lot of time being hypoxic, as can happen if you do frequent airplane travel) can disrupt your gut bacteria and actually harm your mitochondrial system.[12] That's why I label this an advanced hack; you want to approach it thoughtfully.

The Russians put together a complicated system in which pilots would breathe through a series of oxygen-scrubbing sponges. I tried a system like that once. Your lungs fill with air, but you get dizzy at the same time. It's a scary feeling. If you want to try out a hypoxicator, begin gently. Don't go to 85 percent oxygen all day long and stay there, or you'll be in trouble. But if you do it for one minute and then return, then do it for another minute, then return, repeating for a half hour or forty-five minutes, you will send a powerful signal spike to your cells. You definitely want to track your blood oxygen level as you go; you can buy a finger oximeter for less than $20 at a drugstore or online. Monitor your oxygen level until it drops to the mid-eighties. Then breathe normal air until it returns to baseline.

| BLOOD FLOW HACKS |

BIOHACK HIERARCHY

- **Pressure cuffs** are a cheap way to restrict blood flow during exercise but can be risky if you aren't careful.

- **Custom blood flow restriction bands** are a safer and more controlled way to restrict blood flow during exercise to increase growth hormone levels and maximize muscle growth.

Much as you can ramp up your energy and metabolism by limiting your breathing, you can send similar signals by limiting your blood circulation.[13] First you restrict (but do not eliminate!) blood flow to your arms and your legs while you exercise, creating local hypoxia in the limb. You get a sudden, sharp buildup of lactic acid and nitric oxide from your muscle activity. After a short time, you release the blood flow and the buildup of nitric oxide and lactic acid hits your brain all at once. That chemical pulse tells the brain that you exercised more than you actually did. Your body responds, and you see big gains in your energy reserves.

Blood flow restriction also works directly in the muscle itself. Because you are restricting your blood flow, your muscle fibers receive less oxygen than normal. One type of muscle fiber in particular, called slow-twitch type 1 muscle fiber, becomes less active. Meanwhile, the type 2 muscle fibers, which are bigger and faster, actually work better with less oxygen. Type 2 fibers are the ones that enable you to bulk up. When you lower the oxygen level in your bloodstream, your body will preferentially use the muscles that get big faster.

This is another way of tricking your MeatOS. Normally, your body will start by using the more efficient, slow-twitch muscles that are good at sucking up oxygen. Unfortunately, those muscles don't grow very quickly. By limiting their blood supply, you can override your operating system and convince your body to focus on the muscles that do grow quickly.

Something else beneficial happens when you restrict blood flow. Growth hormone levels are 170 percent higher after a blood flow–restricted workout than they would be without the restriction,[14] because the surge of lactic acid convinces the body to release more growth hormone. Blood flow restriction also increases a compound called insulin-like growth factor 1, or IGF-1. IGF-1 increases mammalian target of rapamycin, or mTOR, a potent stimulator of muscle growth. Normally you have to damage muscles by stressing them with heavy weights in order for them to do that. Let's see: You did less exercise, using less weight, for less time, and you got more growth hormone and more muscle growth. You improved your energy and metabolism, and you got a strength and bulk benefit in the process. Kind of a neat hack.

Midlevel Hack: Pressure Cuffs

For us home hackers, blood flow restriction takes remarkably little effort and money. You can get started with basic belts that you wrap around your arms and legs. They don't control pressure very well, but they do work and they cost just $35 or $40. Or you can start out with widely available ACE bandages, if you want to be really cheap. Knee and elbow wraps can work, too, as can cotton elastic bandages—anything that will partially restrict blood flow to the muscles.

The problem with the simple belts and bandages is that it is easy to put them on too tightly and cut off too much blood flow, which is called *full occlusion*. That is really bad for you, because it can lead to dangerous blood clots. If you go this route, be careful not to put the belts on too tightly.[15] You want to restrict blood flow to the veins without occluding your arteries. All of the injuries from the use of belts and bandages come from occluding the arteries. And if you wrap them too tightly, you will get less muscle growth, not more. If you're going to use a cotton ACE bandage, for instance, you want to dial back from the tightest you could wrap it. Call the tightest wrap 10 out of 10, and the loosest 0 out of 10, barely wrapped at all. You want to be at about 7 out of 10, enough to restrict your veins but not your arteries. Remember that each additional wrap adds tightness, even if you don't feel it.

Advanced Hack: Custom Blood Flow Restriction Bands

B Strong belts, designed specifically for blood flow restriction, have inflatable bladders like the one on a blood pressure cuff. I once had Jim Stray-Gundersen, the sports physiologist who founded B Strong, on my podcast.[16] He had spent years working with Yoshiaki Sato, a Japanese bodybuilder who had originated the idea of blood flow restriction as a training technique. Stray-Gundersen wanted to develop a simple, comfortable system that average people could use effectively. B Strong's device looks like a doctor's blood pressure cuff, but it is designed specifically for your thighs and your upper arms. It comes with a little pump bulb and a guide that shows you how much pressure to put on.

With B Strong–style belts, you can set the right level of tightness and then stop inflating it. They are safer and more easily controllable than ACE bandages. But really, you can begin with the ACE bandages that you may already have lying around in your bathroom, wrap both upper biceps—not too tight, not too loose—and do some lifting exercises using less than 50 percent of the weight you normally would. You will get an energy kick and a strength kick at the same time.

You don't have to invest the $300-plus yourself; you can go to a gym or fitness center to use B Strong equipment. At the high end, you could go to a sports trainer who specializes in blood flow restriction. That person will use blood pressure control cuffs similar to the ones you would use at home, but he or she will provide detailed guidance about what exercises to do. I do this at home: I put the blood flow restriction cuffs onto my arms, and then I pick up some weights. I do curls and some dips, maybe a triceps exercise, or I'll double down with some of the other exercise hacks. I'll put them onto my legs, stand on my whole-body vibration plate, and do ten squats.

Pro tip: Blood flow restriction in combination with whole-body vibration will do amazing things to your cells. The day after, you will look ripped and feel energized.

ELECTROMAGNETIC HACKING

For a long time, the medical establishment scoffed at anyone who proposed using magnets to heal and strengthen the body. Electromagnetic hacking had a reputation as woo-woo until a respected orthopedic surgeon named Robert O. Becker documented exactly how much electricity is a part of life. He wrote a classic book that belongs on every biohacker's bookshelf called *The Body Electric: Electromagnetism and the Foundation of Life.* In it, he built his argument around a fundamental aspect of electromagnetism called the *Hall effect,* which states that if a current is flowing through a magnetic field, the magnetic field exerts a transverse force on that flow. We know that electrons carry an electrical flow. We also know that electrons are moving through your body all the time; electric currents are an essential part of cellular metabolism. Therefore, magnets must be able to affect the body by changing its electrical flow.

Researchers keep learning more and more about how extensively electrical flow affects the body. For instance, if you injure your skin or break a bone, it changes electrically. That electrical change creates a signal that tells your body where and how to activate its healing process; and any signal in the body has the potential to be hacked. Until recently, if your leg bone was broken and there was a space between the pieces, doctors would usually just cut off the leg. You were screwed. Now doctors increasingly treat their patients using electrical stimulation therapy:[17] they run a little bit of electrical current on either side of the break. The current flows across the gap and hacks the body's electrical system, telling bone cells to grow where the electricity is.

When you combine electricity with magnetism in the right way, you can manipulate electrical flow without applying electrical current directly. In this way, you can generate a powerful signal for biohacking your energy and metabolism. The magic signal is a pulsed electromagnetic field, or PEMF, in which you expose the body to a magnetic field that switches on and off rapidly, causing corresponding electrical swings inside the body.

Many people find it hard to believe at first that their cells can be affected by a completely invisible electromagnetic field. Then I've watched their eyes widen when I take them to Upgrade Labs and turn on our high-powered pulsed electromagnetic system. I'll put it up against their arm, and their arm will twitch involuntarily, almost as if they're doing curls. What's going on is that the magnetic field in the PEMF device induces an electrical current in the nerve that causes the muscle to fire. At the microscopic level, the pulsed electromagnetic field is opening and closing cell membranes rapidly, up to ten times per second. It is an efficient form of cellular housecleaning that enables the cells to pull more oxygen in and dump out their toxins. With more oxygen, they generate more energy. With fewer toxins, their metabolism runs more smoothly and efficiently.

One of the best measures of your metabolic performance is your bone density: if you have highly dense bones, that indicates that you have enough minerals to build bones and that your metabolic activity is making effective use of those resources to build you up. Sure enough, regular users of high-powered PEMF routinely report having exceptionally high bone density. The only other way you could get so much bone strength and density would be to do excessive amounts of high-impact exercise, which would destroy your joints over time.

People who get regular PEMF treatments also see whole-body metabolism improvements, with healthier and more energetic cells, because their body gets such a targeted, heavy workout signal. High-power PEMF radically drives up the level of bone morphogenic protein in the body.[18] When your bone morphogenic protein level goes up, your cells become better at using glucose and ketones for fuel and your metabolism gets stronger. With more bone morphogenic protein available, your cells are hungrier for energy. They work harder, which tends to help even out any metabolic disturbances.

Advanced Hack: Pulsed EMF (PEMF) Therapy

Unfortunately, there is no cheap and easy way to do PEMF on your own—at least, not yet. PEMF therapy requires expensive, professional equipment like the kind we have at Upgrade Labs. For about $200 you can buy a low-power system that may help with your circulation or hormone levels, but it isn't going to give you big metabolic benefits. For that, you need access to a high-power, professional-level system. That said, the cost of therapy is reasonable, the session is fast (usually twenty minutes or less), and the benefits are considerable, even for people who already do advanced training.

As an example, Nikki Bella, a well-known WWE wrestler and a star of the *Total Bellas* reality show, came to Upgrade Labs one time for us to help her get into shape for her final WWE fight against Ronda Rousey. We used our PEMF system on her neck, where she had an old injury. After the first session, she was startled by the improvement: "Oh, my God, my neck doesn't hurt at all!" The PEMF reactivated the cells, restoring their metabolic power in ways that they thought they couldn't do. Her MeatOS was telling her injured cells to stay safe and dormant. Our PEMF signal showed the cells that they had power, and suddenly they were able to restore flexibility and turn off pain signals that had been active for years.

I'm not just touting my own business here. High-powered PEMF systems are now available at many clinics for trainers to use. You can do an internet search for a facility that offers PEMF therapy near you. Most sessions cost about $50 to $60 per session, but you can often get a discount if you purchase a package deal. You'll want to do multiple sessions to derive benefits from it.[19] Think of PEMF therapy not as an expensive hack but as a relatively affordable way to reduce pain, increase energy, and avoid genuinely costly and often debilitating medical procedures. It's also another very effective way to get the benefits of exercise without wasting your time doing suck workouts in the gym.

9

HACK TARGET:
BRAIN AND NEURO-FITNESS

Before you set out to upgrade your brain, here's a useful tip: your brain is like a gorilla in the forest who has never seen a mirror.

If you don't know what I mean, do a quick web search and look at some of the videos posted online. Scientists have run experiments in which they place a mirror in the wilderness and film what happens when a gorilla comes across it for the first time. The results are fascinating. The animal stops, sees his reflection, and freezes. He thinks for a moment. Then you can almost see the light go on. The gorilla realizes "Oh, wait, that's me!" After that, he will sit there, smile at his reflection, and pick at a leaf stuck between his teeth. Gorillas do not normally have mirrors! That creature never saw his teeth before, yet as soon as he had access to a new perspective on himself, he understood how to make use of it.

Your brain is like that gorilla, primed to operate at a higher level if you just hit it with the right hacks and expose it to the right signals. (To push the metaphor: the sluggishness and fog in your brain are like the leaf stuck between his teeth. You want to get rid of those things, and you can do so easily—if you know how.) Your brain is wired to change rapidly, to optimize itself, and to grow and evolve, more than any other part of your body. Do you want better memory, greater mental speed, and more clarity? Are those your top targets? They are all within your reach.

The problem is that your brain has very little self-awareness, because all of its sensors are looking outward. It is another limitation of your MeatOS, which was honed by evolution for survival, not for allowing you to maximize your personal potential. The brain and the rest of your nervous system has its own type of laziness. It sticks to its job of monitoring the environment around you and doesn't want to waste its resources on self-improvement. That's why you have exactly one feedback system from the brain: the fifth cranial nerve, or trigeminal nerve, which transmits sensations from your face and also controls biting and chewing. Otherwise, your brain pays all of its attention to the world around it and then automatically changes in response to the environment without being able to see itself.

A system that lacks self-awareness is a system that is ripe for hacking. I got my introduction to neurological hacking when I did my first neurofeedback sessions while I was in my twenties. At the time, I was working in Silicon Valley, leading a very cognitively demanding life and seemingly at the top of my game. Yet I was starting to feel stupid a lot of the time. I'd be in a meeting and suddenly forget basic technical details that I should have known off the top of my head. It turned out that I had been exposed to toxins that were interfering with my brain metabolism, but I wasn't aware of that at the time. I just knew I wanted to be sharper and clearer, so I started reading up on neurological hacks.

I bought every book on the brain I could find, read up on all the latest research online about it, and got interested in a then-experimental technique called neurofeedback. The goal of neurofeedback is to make your brain work better by training it to be self-aware, so that it will accept conscious commands that will override the neuro MeatOS—like the gorilla with the mirror.

In my experimental, biohacking frame of mind, I went and bought my own EEG machine, which was quite expensive back then. I started doing neurofeedback on myself, but pretty soon I realized that doing brain work on yourself is dangerous. If you make a mistake, you're not going to know you made it. I realized that it's much better to do neurofeedback with a professional and put my EEG machine into storage. I still have it in a drawer somewhere; I've worked with experts ever since.

There are neurofeedback-based biohacks that you can safely do yourself. The really heavy-duty stuff that can rewire the operating system in your brain falls into a whole other category, however. I eventually opened a clinic, 40 Years of Zen, to develop professionally guided high-end brain performance. There are many ways to activate slope-of-the-curve improvement in your nervous system, but you want to be smart and careful about how you use them.

SUPPLEMENTS TO SUPPORT YOUR NEURO HACK

Creatine

L-theanine

Bacopa monnieri (water hyssop)

Active PQQ

Oxaloacetate

Celastrus paniculatus (black oil plant)

FEEDBACK HACKS

BIOHACK HIERARCHY

• **Home neurofeedback devices** enable you to train your brain to shift to different states.

• **Hemoencephalography (HEG)** uses infrared sensors to measure how much oxygenated blood is flowing through various parts of your brain and is a powerful way to monitor your mental activity so that you can consciously influence it.

• **Expert-guided neurofeedback** with an experienced neurofeedback technician can help you double down on your brain's strengths and address weaknesses.

Accessible Hack: Home Neurofeedback Devices

Neurofeedback is one of the easiest ways to wake up your gorilla brain. You can now buy a consumer neurofeedback system for around $200. One system, called Muse, is a headband that reads your brain wave patterns via a technique called *electroencephalography* (EEG) and sends the results to your phone or laptop. It is designed to teach you how to shift yourself into different mental states. This approach works well for some brains, not so well for others. You need to experiment and see if it works for you.

FocusCalm, another consumer neurofeedback device, also uses EEG to monitor your brain waves, but it combines the readouts with a specific task: it trains you by having you play games on your phone while attempting to maintain calm in your brain. I find this type of feedback really useful because it doesn't just teach you to be calm; it teaches you to be calm while getting things done.

Muse and FocusCalm aren't cheap, but they are accessible enough that most people can afford to experiment with them. If you can try out equipment from a friend or at a fitness center, even better. Neurofeedback is highly personal and variable. Over the years, I've bought neuro gear from about ten different companies and assessed it. I can tell you that different gear does different things.

Midlevel Hack: Hemoencephalography (HEG)

Devices that use EEG are able to detect changes in the brain using tiny electrical sensors, but there are other ways to monitor what is happening inside it. An alternate technology called hemoencephalography, or HEG, uses infrared sensors to measure how much oxygenated blood is flowing through various parts of your brain. It is a powerful way to monitor your mental activity so that you can consciously influence it. You might be surprised by how easy it is to control your blood flow. There's a simple trick that neurofeedback fans sometimes try: they give someone a digital thermometer and tell him to change the temperature of his finger just by thinking.

With just a day or two of mental experimentation, most people can do it. They don't need to understand *how* they do it; they just make it happen.

HEG taps into that same ability and trains you to move blood inside your head consciously. This technique was originally developed to help people like me who have attention deficit hyperactivity disorder, better known as ADHD. It turns out that people like me have low levels of blood flow in the front of our brain when we try to pay attention. Researchers found that they could help people with ADHD improve significantly by using five to ten minutes a day of HEG-based neurofeedback.[1] Patients would play a video game and learn how to move blood in the brain so that the prefrontal cortex, the part that does the conscious thinking, receives more blood, more oxygen, more nutrients, and more energy. Studies have shown that this kind of training can permanently heighten people's attention index,[2] a quantitative measure of their ability to pay attention.

HEG is a lot more accessible than it used to be. A company called Mendi now sells a consumer HEG neurofeedback device that you can buy for less than $400. You put the infrared sensor on your forehead—it looks like a little headband from Tron—and it sends the readings to your phone. Then you do whatever it takes to make a little graphic icon rise on the screen. Studies show that this type of neurofeedback can help treat sleep disorders, attention deficit disorders, and mood and memory issues. It works amazingly well.

Advanced Hack: Expert-Guided Neurofeedback

If you can afford a neurofeedback technician and you have access to a good one in your area, it's worth going for a brain consultation. The process takes about an hour: ten minutes for setup, another ten to twenty minutes to get into a zone, some training, and then an evaluation. It gets expensive if you keep going back to refine your techniques—typically $150 to $200 a session. But a good neurofeedback tech can help you double down on your brain strengths and patch over your weaknesses.

Done right, neurofeedback is orders of magnitude more effective than meditation, because it bypasses words and goes straight into the brain. I recall going to Nepal, sitting in the beautiful Kopan Monastery, and joining a meditation class run by a Swiss nun. She spent a half hour directing us into various uncomfortable poses and telling us to visualize specific images, such as a three-inch-tall gold Buddha. I started to feel ridiculous, and later I understood why: the nun was trying to use words to activate a brain state, but there are just no words to tell your brain, "Turn on occipital alpha waves and turn down the strength of your delta waves, but organize it at the same time."

Neurofeedback is a lot more direct. You can look at the readout and learn how to control your brain state intentionally. You may not know exactly what you are doing, but you can easily sense the result. Then you can learn how to aim for that exact state—not by responding to words but by aiming your brain there directly. I don't believe that any of us has the time to spend two hours a day meditating or ten years in a monastery to get to where we want to go. Neurofeedback is one of the fastest and most important ways to upgrade the neurological MeatOS, so that we can spend a lot less energy and time being jerks to each other. When you have more clarity and self-awareness, you are nicer to yourself and other people. It puts you into the driver's seat.

| DIRECT NERVE HACKS |

Accessible Hack: Vagal Nerve Stimulation

Neurofeedback depends on conscious manipulation of the brain, but there are also ways to influence your nervous system directly, from the outside. Dr. Stephen Porges, a professor of psychiatry at the University of North Carolina at Chapel Hill and the creator of polyvagal theory, realized that stimulating the vagus nerve—the biggest and longest nerve in the body—is a particularly effective way to hack the brain. The vagus nerve runs from your brain through the bottom of your ear, down your neck, and all the way through your gut. It controls your autonomic nervous response, the fight-or-flight response. It is the reason why your stomach turns queasy and you cannot eat when you're afraid.

If the vagal nerve becomes inflamed, it can cause temporomandibular joint (TMJ) disorders,[3] which can produce terrible jaw pain. Porges realized that the vagus nerve is linked with a wide range of other neurological disorders and unwanted brain states as well. Stimulate the vagal nerve correctly, and you can not only tame your TMJ disorder but also relax your whole body in a really profound way.[4] You can instruct it to turn off stress and anxiety, reduce migraines and other forms of pain, and improve your sleep.[5]

Today there are companies that make consumer-grade vagus nerve biohacking devices. One of the companies is called HUSO. (Full disclosure, I am an adviser to it.) It makes a $600 device that combines auditory inputs with small vibrating devices called transducers that you put over your wrists and ankles. The device creates specific modulated frequencies that help create a natural resonance in your body that balances your stress response. When you use the device, you lie there and get profound effects on the vagal nerve and the parasympathetic system, the part that controls the calming "rest-and-digest"[6] aspects of your behavior.

Another company, Apollo, was founded by Dr. David Rabin, a leader in psychedelic therapy at the University of Pittsburgh. The Apollo resembles a watch that you wear. It vibrates on your ankle or your wrist at a frequency that has been shown to give people stress relief and improve sleep. It can even simulate the feeling of a hug—the power of human touch. This touch stimulates the nerve endings that send signals to the vagus nerve, leading to a decrease in heart rate and blood pressure. This also increases the body's level of oxytocin, the hormone that promotes social bonding and sexual pleasure.

A company called Sensate offers a third option for stimulating the vagus nerve. The Sensate device is a little pendant that you place over your sternum. It vibrates at a specific frequency to stimulate your vagal nerve to relax. Its comforting vibrations help you switch on the parasympathetic, rest-and-digest portion of the nervous system, giving you a sense of calm and relaxation.

SOUND HACKS

BIOHACK HIERARCHY

- **Sonic feedback** involves using different music soundtracks to change the state of your brain.

- **Sound therapy** with the Safe and Sound Protocol involves using sound signals to activate the vagus nerve.

- **Sound enhanced with light** involves augmenting sound with colored flashing light to cause positive changes in the brain.

Many of the devices used to stimulate the vagal nerve make use of vibration. All sound waves are a form of vibration, and sound is a powerful tool for hacking your brain states. Many people have a hard time discriminating sounds. I'm one of them. If I go to a crowded restaurant, I can't hear what anyone else is saying; even in a boardroom, if five people are talking, all of my effort goes into picking out their words and meaning. I know that many other people face this challenge as well. It's a hackable problem.

Looking for a solution, I went to an audiologist and did what is called *auditory integration therapy*, or AIT. It's a test used to figure out where the brain is weak at picking up auditory signals. It turned out that I have dog-level hearing at the high end: I can hear the high-pitched sounds that teenagers can hear but adults normally can't, but I have significant gaps in the midrange, around 1,000 Hz.

(The full range of human hearing goes from 20 to 20,000 Hz.) The audiologist then ran a kind of sonic neurofeedback experiment: he played a familiar Mozart symphony with the sound of a cymbal missing in a key place. My brain was straining to hear the sound it expected, but the sound wasn't there.

That simple-seeming hack was enough to strengthen the parts of my brain that had never learned to hear certain frequency ranges. I would listen to different musical brain-training recordings for an hour and afterward feel wiped out from all the concentration. But after I was done, my auditory processing was radically improved. The procedure was life changing. That is the power of sound to tap into your MeatOS.

Accessible Hack: Sonic Feedback

The ears provide a pathway into your brain and into your neural MeatOS. It's an old idea, one that long predates modern technology. The original auditory stimulation was Tibetan "singing bowls." A monk would ring one bowl in one ear, one bowl in the other, and initiate a mental transition state. The brain picks up a sound on one side, a sound on the other, and tries to match them. When it does that, it is forced to depart from baseline. It's a simple, ancient way of pushing the slope of the curve. This is another instance in which singing is a helpful, free therapy: exercising your voice stimulates the vagus nerve[7] and aids recovery from stress.

If you want to experience the modern version of targeted sound stimulation, you can do it essentially for free: go to YouTube, and search for Binaural Beats. There's also Centerpointe, which uses advanced forms of binaural beats; it calls them Holosync sounds. The service costs less than $200. The company's founder, Bill Harris, was an early brain hacker. I've used Centerpointe for several years with good results for making my brain sharper. You listen to audio every night before you go to bed. After about six months your brain feels better ordered—just from the effects of sound.

There's also Focus@Will, a music subscription service that costs about $50 a year. Focus@Will creates instrumental soundtracks designed to induce specific neurological states while you are studying or working out. Different people's brains respond differently to different soundtracks.

Midlevel Hack: Sound Therapy

Dr. Stephen Porges also developed a sound-based nerve-hacking approach called the Safe and Sound Protocol (SSP).[8] It is offered by an increasing number of therapists and involves using sound signals to activate the vagus nerve. You can listen to SSP playlists remotely or in the office of a practitioner using over-the-ear headphones. This technology helps repattern neural networks and takes your nervous system out of a fight-or-flight state into a socially engaged, relaxed state.

It turns out that there are certain frequencies of sound, specifically in the female voice range, that enable you to release and heal neurological trauma. Sound therapy can reset the nervous system and promote recovery from chronic stress. On my podcast, Porges told me a remarkable example of what Safe and Sound Protocol can do. When he was in the United States, he would play the protocol to a room of, say, five hundred people and watch their response. He would always have therapists on hand, because the sound therapy can bring very intense buried trauma to the surface. A few people would usually start crying uncontrollably, just from hearing sounds. His therapists would tend to those people.

Then one time Porges went to London and did the same demonstration to another room of five hundred people, but after a few minutes he had to stop as a lot of the older people in the room had become very distressed. He realized that the older participants had lived through the experience of their city being bombed during World War II. It had created an enormous collective trauma. Many of the younger people in the room were struggling, too; most of them were immigrants who had come to London from war-torn countries. They were dealing with their own systemic, generational traumas. One way or another, about three-fourths of the room was going through a sound-mediated transformation.

I did the Safe and Sound Protocol myself, and I can confirm its effectiveness. You listen to music and go into an altered state with the help of a practitioner. When you come out, you feel like a different person. The practitioners I know who do it also report powerful responses in their patients.

Advanced Hack: Sound Enhanced with Light

Sound is a potent neurological hacking signal, as is light. So why not put them together? Several companies now sell devices that augment sound with colored flashing light. One of the most popular of these devices is BrainTap. It looks like a fancy set of headphones with integrated LED lights that flash over acupressure or acupuncture points on your ears. You can also add goggles that cover your eyes while you listen to audio files that react to the blinking lights and guide you through meditation. It's not cheap, nearly $700, but it creates profound changes in your brain activity. Now you are not relying on words or clanging bowls to influence your neural operating system; you are using guided technology to put your brain into a specific state. The end result is that you can achieve deeper sleep, greater awareness, greater alertness, and greater creativity. Some professional athletes use BrainTap before they go onto the field because it can put their brain into what's called a beta state, with more and faster brain waves.

LIGHT AND VISION HACKS

BIOHACK HIERARCHY

- **Vision exercises** are a simple way to strengthen your vision and help your eyes work together.

- **Reading under red light** improves your eye strength and health.

- **Removing junk light** and shielding your eyes from it can improve the function of your brain and nervous system.

- **Modulated-light LEDs and lasers** can put a signal right through your skull into your brain, stimulating your cells with red or infrared light.

- **LED brain blinking** using rapidly blinking LED lights delivers a much greater dose of light without overheating the eye.

Anything you see creates a signal into your brain, so really any kind of light is a potential hacking tool. You don't need advanced technology. You don't even need an artificial light source; you can also train your eyes and your whole visual system by using some remarkably simple hacks.

Accessible Hack: Vision Exercises

Almost everyone would benefit from this extremely basic technique: Every hour or so during the day, or at least every morning and every night, find a way to look out to a distant horizon. If you work in a cubicle, find a place where you can look out a window; that's enough. Focus on something at least twenty feet away, and hold that gaze for about fifteen seconds. Then toggle back and forth between that distance and something right in front of you, less than a foot away: fifteen seconds on and off, for at least three minutes. What you are doing is changing your eyes' focus from far away to up close. It's free, requires minimal effort, and has significant benefits. If you do this on a regular basis, it will shift your neurology and strengthen your eyes so that you are much less likely to develop vision problems.

Most people have a dominant eye, which means that our eyes are not good at working as a team. Usually it's the right eye, just as most people are right-handed. You can tell by holding your hands in front of you and looking past them so that you see a floaty finger with two fingernails in front of you. When you close your dominant eye, your fingers don't move; when you close your nondominant eye, they do.

Your brain is lazy. It normally uses your eyes one at a time and both of them only when you need stereo vision. But you can hack its laziness and make both eyes work together with a simple, fun hack.[9] Put three beads onto a string about fifteen feet long, arranged with one bead a foot from you, one bead six feet from you, and one bead ten feet from you. Hold the string up to your nose, with the other end tied to a fixed object, such as a doorknob. Sure, you will look like an idiot, but who cares? Focus on one bead, then on the next bead, then on the next, in a cycle. When your eyes start working together, it can improve double vision, headaches, visual discomfort, and other things that suck your energy.

For more advanced training, you can work with a vision specialist, who might use custom video games or VR goggles to improve your stereo vision. When you teach your eyes to work together in concert, the amount of energy required to see goes way down.

Accessible Hack: Reading Under Red Light

In general, doing a lot of close reading undermines your peripheral vision, your stereo vision, and your near/far focus. Studies show that reading underneath red light, even in ten-minute blocks of time, improves eye strength and health. Red light also appears to increase the energy output of the mitochondria in your retina.[10] When I read at night, either I use a red light to read by or I set my phone screen light to red. There is good evidence that reading under red light improves eye health.[11] I use light therapy on my face on a regular basis, and my vision is 20/15 in both eyes.

Light also affects your nervous system by setting or adjusting your circadian rhythm,[12] the central timing process in your brain. Five percent of the cells in your eyes do not collect light for you to see. Instead, they collect light that goes directly into your circadian system. The angle of the light, its intensity, and its color strongly influence the way that it affects your bodily rhythms. One of my companies, TrueDark, has developed glasses that control all of those variables. We have EEG studies indicating that the glasses cause profound changes in brain activation. To me, wearing the glasses for ten minutes creates a brain state that resembles advanced meditation. Doing so has also helped me double my amount of deep sleep.

Accessible Hack: Removing Junk Light

One of the best things you can do to improve your brain and nervous system is to shield it from bad light. What is bad light? Go into any grocery store or big-box store at night, and you will be blasted with high-intensity, blue-tinged LED lights. At the end of your shopping, you'll feel drugged and wonder why you bought so many things you didn't need. As it turns out, those big retailers have studied the biology of lighting patterns and discovered that people buy more when a store is overilluminated.

Your brain evolved to expect angled red light from Mother Nature at sunrise or sunset and overhead bluer light in the middle of the day. The automated systems deep within your body are completely thrown off by strong overhead blue light at night. It activates the suprachiasmatic nucleus, a part of the brain within the hypothalamus that directs the operation of the circadian rhythm. Blue light also suppresses melatonin,[13] the hormone that helps tell your body when to go to sleep, and some forms of blue are even harmful to the retina.

So do yourself a favor: The next time you go shopping at night, put on a baseball hat and light-blocking glasses that stop all four wavelengths that overstimulate your brain (not the commercial blue light blockers, which are ineffective). You will buy less and feel a lot more normal when you come out. Even conventional dark sunglasses can help.

Midlevel Hack: Modulated-Light LEDs and Lasers

We've established that light provides a pathway into the brain, and it turns out that flashing lights in particular have profound neurological effects. Hypnotists have known this for a long time. In 2000, the Norwegian government opened the Lærdal Tunnel, a fifteen-mile-long passage through the mountains that is the longest tunnel of its kind in the world. When the tunnel first opened, the drivers going though had an inordinate number of accidents. People kept falling asleep at the wheel. Safety inspectors realized that the evenly spaced lights above the roadway were exposing drivers to repetitive flashes: blink, blink, blink, blink, blink. That flashing was putting them into an altered state, to the point that sometimes they would lose focus and inadvertently wreck their cars. To fix the problem, the engineers redesigned the spacing of the lights in the tunnel to be random. They also added a couple of break areas where drivers could pull over to walk around and look at the world, because the lighting within the tunnel was so hypnotic.

Modulated light is a powerful biohack, too. Modern LEDs and lasers are so intense and so controllable that they can put a signal right through your skull into your brain to stimulate your cells using red or infrared light. Several companies now sell commercial light-stimulating devices for the brain. They are generally designed as helmets, some using LEDs to deliver the light signal, others using lasers, which are slightly more effective. (There are even people who buy an infrared security illuminator, take off the cover and shine it onto their head. I've tried it myself. It works and it's fairly cheap, about $50, but I don't recommend this DIY approach unless you really know what you are doing.)

My company TrueLight makes one of these light therapy devices. It is called the Baton Rouge, and with it you can shine bright red and infrared light onto selected portions of your brain. Other companies make infrared and red stimulating LEDs that you put up your nose with a little clip so that the light shines onto the base of the brain. All of these devices improve the blood flow in the brain. The ones that you wear outside your head will also stimulate hair growth if they have red light in them. The range of options is vast, from $75 for an LED hair growth stimulator that will have some brain effects to a $5,000 laser cap that triggers intense changes in the brain's blood flow.

The safest devices use LED panels that produce only moderate-intensity light. In these cases, you are unlikely to overdose your brain with light. You can usually run even the higher-power systems for a good twenty minutes. They are designed primarily for your face and hair, although they do also help the brain. If you use a helmet design that focuses light in one spot on your head, I don't recommend doing a dose in any one place for more than a minute. The sign that you're getting too much light stimulation in the brain is that you feel brain fog and sleepiness (a sign that your neurons are worn out) or sugar cravings (a sign that your neurons need more energy). For newcomers, a total of five minutes of light therapy on the brain should be enough. As I've done a lot of light therapy, I usually need twenty-five to thirty minutes to feel fully activated.

Advanced Hack: LED Brain Blinking

The cutting edge of light-based brain hacking uses rapidly blinking LED lights. LEDs can turn on and off almost instantaneously, which gives them a tremendous slope-of-the-curve influence. Advanced brain light manipulation could not be done without LED technology.

The science in this area is just getting started, but a few companies are developing blink lights, either wearable or therapy lights. The goal is to create faster recovery, better sleep, and specific hormonal responses. Neuroscientists and biohackers are just beginning to figure out the best ways to use the light-signaling networks in the body to prompt biological changes. Over the next three to five years, you're going to see a lot of new research come out about this, because light is so biologically powerful and the cost of LEDs is now close to zero. The big question is how to blink them at just the right rate, using just the right colors, to get the desired cognitive and neurological effects. Pulsing the lights rapidly makes it possible to deliver a much greater dose of light without overheating the brain tissues, which opens the door to stronger forms of light hacking.

| ELECTRICAL HACKS |

BIOHACK HIERARCHY

- **Transcranial direct-current stimulation (tCDS)** uses a direct current to stimulate brain cells to make them stronger.
- **Electrical muscle stimulation (EMS)** uses one or more types of electrical stimulation to train muscles or nerves. The best units, such as NeuFit, are multiwaveform, which provides far greater effects.

In the previous chapter I described the ways that electromagnetic signals can improve your cellular metabolism. You can also use directed electrical current to target your brain and nervous system. Essentially, you are bypassing sensory stimuli such as vibration, sound, and light and directly using the language of nerve cells: electricity.

If you put an electric current onto your arm, the electricity will move your arm for you; you are imposing a state versus creating a state. In principle, it's possible to do the same thing to your brain. That's an amazing but frankly terrifying idea: if it is possible to impose a state on your brain, you yourself should be the one to do so. Just imagine if people who wanted to control you could impose a state on your brain. We need to be careful with these technologies. But as biohackers, that's what we do: we explore the frontier; we make open-source systems so that we—and by *we* I include the whole public—can be the ones in control.

This type of research dates back to the 1960s, when Soviet scientists were searching for ways to make the country's astronauts more resilient. Along with their experiments on strength and breathing, they also ran tests to see if they could redesign the astronauts to need less sleep. What they figured out was that they could do it by running a very small alternating current across the brain. They put a little clip on each ear and hooked them to a transcranial alternating-current stimulation, or tACS, device which runs a trickle of current through the brain.

Intermediate Hack: Transcranial Direct-Current Stimulation (tDCS)

The nerves in your body transmit signals via tiny direct-current electrical impulses; the more you use them, the more electricity they have to carry. Like all the other lazy parts of your biology, nerve cells change only when forced. Push a lot of electricity through your nerves, and they build up a thicker layer of myelin, a substance surrounding them that acts like the insulation on an electrical cord. In technical terms, they become more heavily myelinated. If there's a small amount of current running between your ears, the cells are going to myelinate better because they need to carry more electricity. The result will be that they can move electricity much more quickly. Neurologists have a test called a nerve conduction study, which measures the speed at which a nerve can take electricity. If you do electrical training on a regular basis, you develop a super-powered nervous system that can carry electricity faster and with less effort than most other people's can.

Today, you can buy tACS devices starting at about $500 and running to thousands of dollars for clinical-grade machines. I own one of the clinical-grade devices to help me with my writing. It is a profound form of exercise for the brain. tACS has mostly fallen out of favor and been replaced by tDCS, a similar technology that uses direct current instead of alternating current.

In the original days of electrical brain stimulation, when I was first starting to blog about it, there were no commercial companies making equipment to do tDCS. We biohackers had to buy a prescription device designed for forcing drugs through the skin with a small electric current system called *iontophoresis*. I got one of those machines. I would hook up one electrode to a specific part of the brain (you really want to get the placement right!) and then attach the other electrode somewhere on my body. We were operating by trial and error back then, but we know now that you want to target the front of the forehead, especially the left prefrontal cortex, for the electrical input to have an effect.

With the modern consumer-grade devices, there is very little risk. You can buy one for as little as $150. It may seem like exotic technology, but it's actually low-hanging fruit in terms of biohacking benefits. You soak little pads in salt water, install them on a headset, put the headset onto your head, and turn it on. Boom, you are stimulating your brain's attention center. Both tDCS and tACS also increase neuroplasticity, the ability of the brain to adapt and rewire itself, and increase levels of brain-derived neurotrophic factor, which promotes the growth of new neurons. If you have a brain that's completely fogged up, as mine used to be, you can stimulate the front of your brain for twenty or thirty minutes, which will strengthen the part of the brain used to pay attention. If you do this every other night for a month, you're going to give your brain metabolism a huge lift and you'll feel sharper.

Dr. Daniel Amen, a leading brain scientist, told me to get a Ping-Pong table and play the game as a way to test for improvements in my cognitive performance. I could really see the change. I now play against my twelve-year-old son. He can absolutely kick my ass unless I break out my tDCS system and stimulate my motor cortex. As soon as I'm done, it seems as though the ball slows down and I can keep up with him. He'll actually complain, "Daddy, could you go stimulate your brain? I'm having to slow down for you." If I can keep up with the reflexes and neuroplasticity of a twelve-year-old, my brain must be using its electricity well.

Advanced Hack: Electrical Muscle Stimulation (EMS)

What could you do if you wanted to be crazy about pushing electrical stimulation to the limit? Well, there are clinical-grade machines that are unparalleled. You could go to a sports trainer or a physical therapist who has this kind of dual-waveform electrical stimulation. There are even a few fitness studios opening up that do this kind of training, but they require you to wear sweatpants, spray yourself with water, and put on electrical stimulation shorts and a vest. The procedure is intense, but you get a nervous system upgrade and a muscle upgrade that is unparalleled.

I have this kind of gear in my office, because of course I do. It is a device called the NeuFit, which was invented by an engineer and neuroscientist named Garrett Salpeter. When I crank it up and hit my gorilla brain with a jolt of electricity, the benefits are profound. Is this kind of advanced hack right for you? First you want to know the limits of what your nervous system can handle. Then you can decide how much work you are willing to do, knowing that the work you put in now will pay off in needing to do less work once you are thinking more clearly. Above all, you want to know your targets. You can't do everything all at once, and that's just fine.

10

HACK TARGET:
RESILIENCE AND RECOVERY

If you want to master the laziness built into your MeatOS, you have to push yourself to higher energy levels—that's been my consistent message throughout this book. For improved strength, cardiovascular fitness, energy, or brain function, the routine is fundamentally the same: apply intense input signals that tell the body it needs to perform beyond its normal boundaries of operation, and do it in a steep, slope-of-the-curve style that tells the body it needs to be able to switch rapidly between full-throttle action and comfortable, baseline cruising. Once your biology knows what it needs to deal with, it will adapt to handle it—so that *you* can handle it.

But what if you don't have a good baseline to start with? Many people carry so much stress that they are constantly blowing through more energy than they should, wasting it on anxious feelings and undirected behaviors. It's like going to the gym and doing endless low-level exercises that make you feel miserable and tired but never improve your strength. Those people don't need to focus on pushing their energy output high above baseline, at least not at first; they need to start with biohacks that will help them lower their baseline stress.

Are you one of those people? Then you may want to consider making stress your top target for improvement. If you carry too much stress, don't go for a workout; go for a detox. You have to remove stressors and tell your body to calm down before you can build it up. Fortunately, stress management is itself hackable. You can get to a calm, relaxed state a lot faster and more easily than you may have thought possible. Then you will be in a much better place to carry out the other techniques that will train you to exercise and recover quickly. The response spike in the body does a lot more work in a calm person than it does in someone who is stressed out, stuck unproductively near the peak of the energy curve. All of the energy hacks build on one another.

Keep in mind, too, that you must start with good resources no matter what aspect of your life you are targeting. For destressing and detoxing, your body requires sufficient minerals (including trace minerals), fats, vitamins, and amino acids. Those raw materials are essential to any fast, efficient biological response, even if the response you are pursuing is one of calm rather than one of strength. It's just like building a house: if you don't have the right resources, nothing is going to happen. You have to pull together the right supplies of plywood, Sheetrock, wiring, plumbing materials, and so on. It doesn't matter how quickly you want to see that house go up; without the essential supplies, your house—like you—is going to sit there unfinished.

SUPPLEMENTS TO SUPPORT YOUR DE-STRESSING AND RECOVERY HACK

Rhodiola rosea

Ashwagandha

Holy basil

Ginseng

L-tyrosine

L-theanine

Magnesium

SLEEP HACKS

BIOHACK HIERARCHY

- **Maintaining good sleep hygiene** by keeping your room at a cooler temperature, blocking junk light, and avoiding large meals near bedtime can help you get better sleep.

- **Protecting your breathing** by using a bite guard and/or taping your mouth closed at night will encourage healthy nasal breathing.

- **Tracking your sleep** with the SleepSpace app can help you find out more about your sleep quality and whether or not your sleep hacks are working.

The number one way to relax and recover is boring and familiar, but also free and extremely effective. It's called sleep. When you sleep, your body releases a cluster of signaling substances—including human growth hormone and testosterone, in both men and women—that instruct your cells to repair themselves and regrow. When you do a physiological task that puts a strong signal into the body, such as heavy exercise, it triggers an inflammation response that signals your body to repair itself and grow stronger. Your lazy body then is forced to act, initiate a recovery response, and build new muscles, tissues, and mitochondria. When you sleep well, you produce more of the recovery signals and rebuild more quickly.

Anyone who does heavy lifting or who exercises really hard will tell you that you need more and deeper sleep the night after a hard workout. You're tired after you exercise because anything that stresses the body, including nervous system stress, requires a cleanup and recovery process. The quickest way to get more deep sleep is to do a CrossFit workout. I promise you, you'll get twice as much deep sleep as normal afterward, because you beat the crap out of yourself. But I'd prefer that you not beat yourself up, and I'm sure you would prefer that as well. What everyone wants, bottom line, is to be his or her best self.

Most of us walk around with too much neurological stress and anxiety. If you have tissue stress from physiological factors on top of that—from exercise, tension, or both—it is too much to handle. A lot of us even develop anxiety about anxiety and stress about stress. You can escape this trap by prioritizing sleep as a part of your recovery process. Learning to hack your natural sleep systems will help you reduce stress and get you to where you want to be more quickly. As with all hacks, the process can be surprisingly simple once you know how to do it.

Starter Hack: Maintaining Good Sleep Hygiene

Getting high-quality sleep is the most effective stress recovery technology. It will help you with all your other targets as well. I know it seems simple, but no matter how many times I repeat this advice, people need to hear it again. It's really hard to maintain good sleep habits, especially in our stimulus-oversaturated world. I've therefore created a cheat sheet for you.

Here are the fundamentals of good sleep hygiene:[1] Dim the lights at night. Use blackout curtains. Set your thermostat to 60 to 67 degrees F. Wear junk light–blocking glasses after the sun goes down (the real kind, not the $10 kind you find online). Shut down your electronic devices a full two hours before you go to bed. Increase the warm light setting on your phone, so you get more sunset-simulating red light. Stop eating at least two to three hours before bedtime. On top of those basic techniques, I recommend daily meditation. Over time, it will rewire your brain, strengthening the neural pathways that reduce stress.

Starter Hack: Protecting Your Breathing

Your brain requires a generous supply of oxygen while you sleep so it can do its work of guiding the body to recover. Oxygen is necessary for making growth hormone and testosterone and for the overall operation of the hypothalamic-pituitary-adrenal axis, a complex system of hormonal signals and feedback that maintain homeostasis. People who snore or suffer from sleep apnea don't get enough oxygen at night, which retards all of those restorative systems.

The number one thing you can do to improve your breathing at night is to get a bite guard. Many people grind their teeth and clench their jaw when they sleep, which squeezes the muscles in the jaw along the trigeminal nerve, which affects the blood flow to the brain. Often people have no idea they are doing this; they just know that they feel sluggish in the morning. Many people also experience airway obstruction while they are sleeping, leading to snoring and sleep apnea. Add a little cushioning to your jaw, and it will help position your mouth to ensure that your sleeping brain is getting the oxygen it needs. A cheap drugstore bite guard is effective enough to make a big difference.

Here's another magical, simple sleep hack. Put a small piece of porous sleep tape (widely available in drugstores or online) over the front of your lips to hold your lips together when you sleep. The technique, called mouth taping, forces you to breathe through your nose during the night, which sends more oxygen to the brain and increases your nitric oxide levels. I have been doing it for years. My daughter started doing it when she was thirteen and saw immediate sleep improvements. You will get more deep sleep, snore much less, and wake up better rested. As a bonus, your jaw will form better, your teeth will grow straighter, you won't get as many cavities, and you won't get morning breath. Yes, you'll look a bit silly with tape on your mouth, but the results will probably improve your life.

Starter Hack: Tracking Your Sleep

You can monitor the effects of your sleep hacks by using an app on your phone; it may already be preloaded for you. For a more advanced approach, you can use SleepSpace, a sleep app and subscription service developed by a cognitive scientist named Dr. Dan Gartenberg. He tested and validated the SleepSpace technology with a multimillion-dollar age study.[2] The app combines sleep tracking with sound therapy to improve the amount of deep sleep you get.

Sleep trackers are great for evaluating all of the other sleep hacks. Every night, before you go to sleep, write down quick notes about how you feel. I've been doing this for years. When you wake up in the morning, take your readings with whatever kind of sleep-tracking device you're using. If you also use a broader health tracker, it will measure your heart rate variability or give you a "readiness score" based on your data. But the most important measure is a subjective one: Did you wake up feeling full of energy? If you're feeling well rested and you have more energy throughout the day, that's a pretty good sign that your sleep hacks are working.

| LIGHT HACKS |

BIOHACK HIERARCHY

- **Sun therapy** is a free way to get exposure to infrared and red light.
- **Red-light therapy** using special red- and infrared-light devices can help speed recovery and reduce inflammation.

As I've discussed, the use of the proper light can be an effective way to crank up your cognitive power. You can also apply light in a different way, to reduce stress and put your brain into a restorative state.

Light activates an enzyme called *cytochrome c oxidase*, which causes the mitochondria to manufacture more ATP and generate more energy. Your mitochondria extract energy from fat or sugar by combining it with air to make an electron. The electron moves along a chain of molecules until the energy from the electron is harvested by cytochrome c oxidase. (By the way, this process requires three copper ions and two iron ions, along with zinc and magnesium. You absolutely, positively need your minerals in order to generate full energy.)

Cytochrome c oxidase is a chromophore,[3] a color-producing molecule that strongly absorbs specific wavelengths (colors) of light. In this case, the molecule absorbs light between the mid–600 nanometer (nm) to mid–800 nm range, which is red to infrared light, and can use it to make additional electrons that aren't from ATP.[4] The light enables cytochrome c oxidase to make energy more effectively, so your cells work better. It's like giving your body an extra power cell that you can use for regeneration. As a result, chronic inflammation and stress levels are reduced. As a bonus, this type of light also enhances collagen production, improves fine lines and wrinkles, and speeds wound healing.[5]

Starter Hack: Sun Therapy

The cheapest available source of amber, red, and near-infrared light is . . . sunlight. Full-spectrum sunlight also stimulates the body to produce vitamin D and triggers the production of serotonin, which can elevate your mood. Serotonin is a precursor of melatonin, a hormone important for sleep, so morning sunlight exposure can help with melatonin production when nighttime approaches. In addition, we in the biohacking community know that getting sunlight on your testes increases testosterone production.[6] At the same time, red light releases nitric oxide, which is inherent to having an erection. It works well over nipples, too. If you don't believe me, try a little nude sunning and watch what happens the next morning. Biohacking is all about experimentation, right?

Step one is to go out into the sun. Late morning to early afternoon is best to get the full benefits. It doesn't have to be over your reproductive organs, but it can be. I recommend getting fifteen to twenty minutes of direct sunlight exposure on your skin daily. The lighter your skin, the faster you'll get burned. If at twenty minutes you are getting too red, dial back the dosage.

Starter to Advanced Hack: Red-Light Therapy

If your budget is tight but you want more concentrated light therapy, you can buy an infrared illuminator (available for less than $20 on Amazon) and take off the lens. It produces the 660 nm light that activates cytochrome c oxidase, and the output is strong enough to produce a useful recovery and destressing effect. You can shine this onto your skin or over sore muscles for a reduction in inflammation.

A level up, you can buy small LED panels, about twelve or fourteen inches on a side, that contain an array of red, infrared, and amber lights. You can use them on specific areas of your body that need healing. LEDs also come in wearable form as belts or headsets. For greater intensity, you can buy much larger red-light panels, which are usually made of several smaller panels stuck together. These panels will cost you anything from $30 all the way up to $1,000 or more. If you are serious about your light therapy, you can go to a doctor or a chiropractor to use high-power, clinical-grade therapy lights that cost a few thousand dollars. There are even whole-body light beds available at some medical spas and fitness centers.

It doesn't take much exposure to reap the benefits of light.[7] Various studies have shown that three minutes a day is all you need.[8] For red-, near-infrared-, and yellow-light therapy, I recommend twenty to thirty minutes per day over each target area. Light therapy devices vary so much that you need to experiment to see what works best for you. It's also important to pay attention to the total dose of light you receive. Some of the brighter LEDs are so powerful that they can burn your skin.

| HERBAL HACKS |

Stress manifests itself through chemical changes in the body, so why not go after it by chemical means? That's the idea behind adaptogens, a term introduced by a Soviet toxicologist named Nikolai Vasilyevich Lazarev in the 1950s to describe substances that induce a "state of nonspecific resistance" to stress.[9] He was onto something, though it took a while for the rest of the world to catch up with him. Researchers and healers have now identified a whole family of adaptogenic herbs that can help you go into and out of stress responses faster than normal.

The top three adaptogens are ashwagandha, ginseng, and rhodiola. They reregulate the endocrine, nervous, immune, digestive, and cardiovascular systems. They all meet the three criteria that define a "primary" or full-function adaptogen: they help resist stress, they help maintain physiological equilibrium (homeostasis), and they cause no harm or notable side effects. There are many secondary adaptogens as well. Among these, I like eleuthero, sometimes called Siberian ginseng, which suppresses stress and amplifies immune function. My very favorite one is holy basil (*Ocimum tenuiflorum*), a peppery-tasting plant used in traditional Ayurvedic medicine. In addition to its antistress and antianxiety properties, it is believed to help with high cholesterol and depression.

Adaptogens are safe for most people to use on a regular basis, although you might want to increase your doses when you're under a lot of physiological or psychological stress. Most of them are available as tinctures, in capsule form, or even in herbal tea form. You can refer back to chapter 4 of this book for more specific information on dosing. Of course, if you're taking any medications, it's best to talk to your doctor first.

HEAT AND COLD HACKS

BIOHACK SUMMARY

- **Sauna therapy** gives you exposure to extreme heat, which tells your mitochondria to grow stronger.
- **Infrared sauna therapy** gives you the benefits of regular sauna therapy with the bonus of infrared wavelengths to enhance healing and relaxation.

- **Ice baths** for fifteen to twenty minutes are an uncomfortable yet effective way to get a strong biological response.

- **Cryotherapy** is a more comfortable form of cold exposure in which your body is exposed to vapor from liquid nitrogen or another ultracold source—typically −270 degrees F.—for three minutes.

- **Controlled breathing with cold therapy** helps train your body to stay calm during stressful situations.

Heat and cold are powerful signals for reducing stress and inducing detox in the body. Both heat and cold are dual-purpose slope-of-the-curve treatments. They are stimuli that create a response, and that response, in turn, helps you recover more quickly. By using hot and cold therapy, you will become one of the resilient people whose MeatOS can handle whatever life brings your way. Then after you finish the recovery part of the process, you will be able to stand up, dust yourself off, and do something else—whatever is important to you.

Ironically, you need a strong energy supply for your body to know that it is safe to lower your baseline stress. Only then can you be truly calm and destressed. Heat and cold exert a tremendous influence on the organelles at the center of your body's energy supply, your mitochondria. While they are cranking out energy, the mitochondria are also shedding a lot of thermal energy, or heat. That heat keeps us warm and keeps critical biological enzymes operating correctly. The nature of your enzymes is that they work only within a very narrow range of temperatures. If your body temperature drops to 97 degrees F. or below, many of your enzymes simply won't function. Mitochondria also need body heat in order to purify water in the body so that it can transport electrons.

In other words, if you get too hot or too cold, your whole energy system is thrown off. And your MeatOS knows it, so it does everything in its power to fix it. If you expose your body to extreme cold, your MeatOS registers an emergency. It thinks, "Oh, geez, I might die and never have babies," so it turns on the capability to become really warm really quickly.

You didn't require the ability to turn the heater up to full blast before, because it never mattered. This temporary stress actually helps your MeatOS upgrade itself in the long run. Now that your body knows it matters, it will reregulate your metabolism. It will start sucking up all the available sugar and make sure you are properly sensitive to insulin. It will also start adding beige and brown fat, the healthy kinds that generate a lot of heat. They produce another type of fat, cardiolipin, that is essential for building the membranes of your mitochondria. Beige and brown fat runs all up and down your spinal cord and along your neck. Kids have a ton of it, but adults have very little of it—unless they do cold therapy. According to Andrew Huberman, a neuroscientist at Stanford University, cooling specific body sites can increase endurance and strength output by 200 to 600 percent,[10] prevent muscle soreness, reduce brain fog, and enhance recovery by stimulating adrenaline.

Heat and cold also influence your energy supply through PGC-1α, the compound that is activated when you exercise. You may recall that PGC-1α stimulates mitochondrial biogenesis, the creation of new mitochondria. In a recent study, a group of Korean researchers put one group of mice into cold water, right above freezing, and compared them to another group of mice that had to swim at a normal temperature. After eight weeks, the exposure to cold had increased the PGC-1α levels in the mice far more than the exercise had. When the mice exercised and were exposed to cold exposure together, they got the best results of all, significantly better than exercise alone.[11]

Heat does good things for your mitochondria, too, by increasing their consumption of oxygen, which improves their efficiency.[12] Heat also helps kill off weak, inefficient mitochondria.[13] In short, cold therapy increases the number of mitochondria while heat therapy makes them more efficient, and both cold and heat help kill off the weak ones.[14]

An important note: Both heat and cold send a signal that makes the body thinks it has exercised, causing the body to dump a bunch of toxins. That's very good for you in the long run, but it also means it is crucial to get rid of those loose toxins. It is best to combine heat therapy and cold therapy with a detox, such as the ones I describe later in this chapter.

Starter to Midlevel Heat Hack: Sauna Therapy

Studies have shown that the regular use of saunas can make you live longer, help you recover faster, and promote detox from sweating.[15] They mimic exercise in some ways, so they reduce cardiovascular risk. Personally, I enjoy my time in the sauna, because I can watch a show, listen to an audiobook, or talk to someone while I'm locked away from the rest of the world. Or I can just peacefully meditate.

Three times a week of sauna is enough to be effective. Studies from Finland, Sweden, and Norway all support using a sauna three to five times a week for a minimum of twenty minutes at a time.[16] I usually need forty-five minutes to get up a good sweat, probably because I'm used to it. As a general rule, you want to start short, around five to ten minutes, and make sure you pay attention to your level of sensitivity.

You probably don't have room for a giant sauna in your house or apartment. Fortunately, you have options. You could have a far-infrared sauna with some near-infrared lamps in it if you want to be ambitious about it. You'll learn a bit more about infrared saunas, a more advanced hack, in the next section. You could get a sauna blanket that wraps around you, which is a whole lot cheaper; it's like a sleeping bag that has infrared heaters inside. That'll save you money and space. Or you could just use whatever's available in your area: go to a steam sauna, go to a dry sauna, go to the sauna in your gym. Just get hot.

For an extremely economical sauna experience, take an insanely hot bath. Did you ever see one of those old cartoons where somebody puts his feet into a bucket of hot water? It actually works. One time I got very chilled in a hotel, so I soaked my feet in a bucket of superhot water. Within ten minutes my whole body was sweating. You can raise your body temperature quite a bit that way. Raising your body temperature is like having a brief fever; it strengthens your mitochondria and enhances the function of your enzymes. Elevated body temperature also likely produces antiviral or antibacterial effects because viruses and bacteria generally don't do well at high temperatures. That's one of the reasons your body raises its temperature when you have an infection.

Midlevel Heat Hack: Infrared Sauna Therapy

Infrared saunas also give you a mitochondrial lift, but they exert their positive benefits by heating you from the inside out. Full-spectrum infrared saunas provide near-, mid- and far-infrared wavelengths and heat your body's core to a cellular level, where most toxins are stored. Infrared's deep penetrating heat stimulates metabolic activity and triggers the release of stored toxins through your sweat. Other benefits of full-spectrum infrared sauna therapy include lowered blood pressure, relaxation, improved circulation, pain relief, accelerated wound healing, weight loss, and increased tissue oxygenation. You can purchase an infrared sauna for your home, visit Upgrade Labs, or go to a gym that has an infrared sauna to feel the benefits for yourself.

I use a Sunlighten mPulse 3 in 1 sauna, which gives me the full spectrum of infrared wavelengths so I can get all of the benefits. After spending twenty to thirty minutes inside, I come out feeling completely relaxed and rejuvenated.

Don't be a perfectionist about heat therapy. If you do three to five sessions a week, that's great, but even once a week has benefits. Aim for sessions of fifteen or more minutes, but just do the best you can. I sometimes do infrared. I also have a steam sauna, and sometimes I'll do that. The key thing is to raise your body temperature several times a week and perhaps work up a good sweat to detox.

Getting into a sauna can help you add oxygen to sore muscles and lose weight. Just keep in mind that you have to recover from the sauna—recover from your recovery, in effect—because you liberated a lot of toxins. Heat is part of the recovery process, but it is not a complete recovery in itself. You will need sleep and maybe a little bit of vibration or lymphatic drainage (see page 211).

Starter to Advanced Cold Hack: Ice Baths and Cryotherapy

On the cold side of things, the big question is figuring out the minimum effective dose, because you've got other stuff to do, and frankly, cold therapy is not a lot of fun.

At the low end, you can dip your whole body into ice water. Buy a bag of ice, put it into a big tub full of cold water, and get in. The problem with that approach is that it's not very efficient. Did you ever try to fill a whole tub with ice? You could get more creative and go to Costco, buy a chest freezer, and fill it with ice water. That's relatively inexpensive, and it works. It's just not pretty or especially convenient. Get into an ice bath for fifteen to twenty minutes, if you can handle it, and you will get a strong response. It is bone-chillingly cold, totally different from a wearing-shorts-in-winter kind of cold. More pain, more slope-of-the-curve signal.

If you want fast results and a more comfortable form of cold exposure, you can try cryotherapy,[17] a kind of treatment in which the body is exposed to vapor from liquid nitrogen or another ultracold source—typically –270 degrees F.—for three minutes. You will get a large endorphin rush when you're done. Cryotherapy has become popular lately, so it should be relatively easy to find a treatment center near you. You can expect to pay around $50 to $60 for a whole-body session. It is very cold, meaning that the slope of the curve is high, but the superchilled vapor doesn't pull your whole-body temperature down the way an ice bath does. Instead, it mostly hits your peripheral temperature receptors, the ones on your skin.

To create a home ice bath for your face, you can buy a mini–cold therapy device that's basically a superchilled salad bowl as big as your head. You put an inch of water into the bottom and put the bowl into your freezer. The next day, before bed, you add some cold tap water and stir it a little bit so that the ice that's in the bottom melts into the water. Now you have really, really cold water.

You take a deep breath and stick your face in there, like a dippy bird, and hold it in there as long as you can—maybe eight to ten seconds. Then you emerge, take a breath, repeat, and do it several times. Or at that point you could do what I did: buy a snorkel and stick your face into a bowl of ice water or an ice-filled sink for two minutes. The greatest concentration of temperature receptors in the body is in your chest and face, so this approach is actually quite effective. Simplest of all, just stand in the shower and do one minute of cold water hitting your face and chest. Most people can't do one minute.

One study showed that after three days of cold exposure, the level of cardiolipin in the mitochondrial membranes increases.[18] This is why, at minimum, you probably do want to take a cold shower with the water directed onto your face and chest for as long as you can. On day one, you'll make it to twenty seconds if you're lucky. On day two, you might get to thirty or forty seconds. On day three, you might survive a minute, but you'll still be swearing the whole time about how miserable it is. Then, on day four, things will magically change because your mitochondrial membranes will have changed. You'll be able to go many minutes, and you'll feel better and invigorated.

Starter Cold Hack: Controlled Breathing with Cold Therapy

Breathing exercises enhance the effects of many of the biohacks described in this book. The rapid, hyperoxic/hypoxic style of breathing developed by Wim Hof (see chapter 8) is especially useful for getting the most out of cold therapy.

But the most important breathing style for reducing stress and resetting the nervous system is the "box breath": you inhale for five seconds through the nose, hold for five seconds, exhale for five seconds through the nose, then hold your lungs empty for five seconds and repeat. It's just a box, five seconds per side, however many times you need. It'll take you out of fight-or-flight mode and put your nervous system into rest-and-reset mode. Box breathing is widely used by special forces officers to remain calm and focused. Another breathing technique that's important for destressing and recovery is ujjayi, a type of controlled yogic breathing. Combining calming breathing techniques with intense cold therapy can help you train your body to stay calm during stressful situations.

If you carry a lot of stress, it keeps you at a constant low boil of energy expenditure, as if you are constantly working out at a medium, inefficient level. Try lying on the floor on your back so that your blood pressure can return to baseline more quickly, and do box breaths through your nose to rapidly get yourself into chill mode. This will help you reset if you just did a high-impact workout. Just as remarkably, it will work if you are a bundle of nerves as a result of pressures at your job. It's a superpower you can summon on demand, because it's latent in your body's operating system.

| DETOX HACKS |

BIOHACK SUMMARY

- **Detox your lymphatic system** with exercise, whole-body vibration, a lymphatic massage, or compression therapy.

- **Boost your liver function** with glutathione, calcium-D-glucarate, glycine, and electrolytes.

- **Improve your kidney function** by drinking an adequate amount of water and getting enough magnesium.

- **Cleanse your gut** with toxin-binding substances such as activated charcoal and humic and fulvic acids.

- **Clean up your environment** by managing your stress, getting out of toxic situations, and identifying and removing any chemicals and mycotoxins that may be surrounding you.

Cold therapy alone is effective, heat therapy alone is effective, but they work even better when you combine the two: a sauna and an ice bath, for instance. Susanna Søberg, a Danish metabolism expert, argues that if you switch back and forth between the two therapies, you should always end on cold, because that forces your body to use energy to heat up.[19] Cold exposure turns up your metabolic function and causes the release of cold-shock proteins,[20] small proteins that can bind to DNA and RNA in the body. Cold-shock proteins dial up your metabolic function and cause your mitochondria to rejuvenate themselves, because the ones that can't make heat quickly are replaced by ones that can. Heat exposure releases a different family of proteins called, naturally, heat-shock proteins. They protect other proteins from damage and help refold misfolded proteins. Basically, they are all-around guardians against stress and molecular injury.

These benefits come with a huge note of caution: cold and heat initially place an additional load on your resilience. When you introduce small but intense stressors, you are also activating mitochondrial and cellular autophagy. In other words, you are blowing up your old, inefficient cells. What happens to the wreckage? Some parts of the old cells will be broken down to make building blocks for new mitochondria. But a lot of those parts will end up as waste products and inflammatory chemicals in the bloodstream. They will have to be removed by the liver and kidneys, which is why hot and cold treatments give you the best results when paired with a recovery process of their own, specifically, a detox.

Your body burns hundreds more calories than usual if you do cryotherapy, especially after a sauna. You've got your heat-shock proteins. You've got your cold-shock proteins. Your body is saying: I was burning more calories from the sauna; now I'm freezing, and I have to warm myself up. I'm going to burn even more energy. Your energy demand is going up, and your calorie intake needs to go up as well. It is crucial to stay properly fueled and clean up the toxic mess so that you can get the proper benefits of hot and cold. Showering will help get rid of the toxins that you sweat out in the sauna. What you really need to do, though, is a deep detox that will help you get rid of the toxins that were released into your bloodstream that you didn't sweat out.

Hack: Lymphatic Drainage

When you see a construction site, there are always dumpsters full of construction debris nearby. Your body's not so different. It has three main systems for getting rid of toxins, its construction debris: the lymphatic system, the liver, and the kidneys.

The lymphatic system's main job is to remove cellular waste and dead red blood cells, pushing it through the lymph—the fluid that fills the body between all of its cells and tissues. The system doesn't have its own pumps. Instead, it relies on body movements to push lymph through the tissues, into the lymph nodes, and into the spleen, ultimately discarding unwanted chemicals into the liver. About 20 liters of blood plasma flow through your arteries. After a day, 17 liters are returned to circulation, while the remaining 3 liters leak out into the lymphatic system.

Humans are not particularly good at detoxing compared to other animals. If you want to get your lymph moving, you need to give it some help. You can get a lymphatic drainage massage; you can get onto a rebounder and jump up and down to get your body moving; or you can do whole-body vibration on a vibration plate. Weight lifting, interval training, or even a moderate-paced walk can all help, too—just get your body moving. If you want to be really fancy about it, you can wear compression pants, which create pressure that drives lymphatic circulation. You can purchase compression clothing at most sporting goods stores.

To get toxins out of the body more quickly, you want to increase the speed of lymphatic flow, which is something you will do with movement. A well-targeted lymphatic massage accelerates the rate of lymphatic drainage. A lymphatic massage is nothing like a Swedish massage. A hard massage causes the lymph system to close off. A proper lymphatic drainage massage feels as though you didn't get a massage at all; it's as though someone's petting you or brushing your skin. The lymphatic system is responsive to light touch, not heavy touch. You can easily do a lymphatic drainage massage yourself on your face. There are diagrams readily available online to guide you. You gently brush your face with your fingers—some people use jade rollers—moving outward from the eyes and nose. Those motions reduce puffiness in your face.

At some wellness centers you can also find clinical-grade compression sleeves, which are a faster and more effective means of draining lymph. I have some at home, so I will lie down and zip myself into a suit that goes all the way up almost to the middle of my chest, right below my rib cage. Then I switch it on, and it slowly, rhythmically moves lymph from the toes of my feet up to where the lymph moves into the subclavian vein near the heart.

Hack: Boost Your Liver Function

The body's second major detox organ is the liver. The liver carries out its detox processes through two main enzyme system pathways, known as phase I and phase II. Most toxins are removed or neutralized by these pathways, so it's essential to have the vitamins, minerals, and other nutrients to support them. Specifically, they require the important compounds glutathione, glucaric acid, and glycine.

You can increase your body's production of glutathione by eating raw, undenatured grass-fed whey protein, or you can get more by taking oral glutathione as a supplement. You can also take a glutathione precursor—alpha-lipoic acid, L-glutamine, or N-acetylcysteine—which can be turned into glutathione by your liver. If you're going to take a glutathione supplement, I recommend taking it in liposomal form, which is better absorbed and utilized by your body. You also don't want to take glutathione every day to ensure that your body doesn't reduce its own production.

Glucaric acid is found in oranges, apples, Brussels sprouts, broccoli, and cabbage; you can also get more of it by taking a calcium-D-glucarate supplement, which removes synthetic estrogens from your bloodstream. You can easily find this in capsule form.

Glycine is the most common amino acid in collagen, so you can increase your intake with collagen supplements or with bone broth, which is also a good source of glycine. You can mix a scoop of collagen protein into your morning coffee if you aren't fasting. Getting more electrolytes, such as potassium, sodium, and magnesium, can help with the liver detoxification process as well.

If you are looking to empty your biological trash dumpsters more quickly, you need all of your liver pathways to be working efficiently. Then your body will be able to sweep away the built-up toxins that are released by hot and cold therapy.

Hack: Boost Your Kidney Function

The third major detox organ is the kidneys, in humans much more than in other animals. Your kidneys play a role in regulating your electrolyte and fluid levels. They also filter your blood and send waste products and excess fluid to your urine so you can safely excrete them. Having too much calcium in your diet relative to magnesium overloads the kidneys. To keep a lid on stress and promote detox, you can increase your intake of magnesium. Get at least 500 to 1,000 mg of magnesium per day. It also helps a lot just to drink enough water and to get a sufficient supply of all electrolytes, the critical electricity-carrying minerals. Even though too much calcium in the diet can overload your kidneys, a special compound called calcium AEP protects the kidneys from damage and helps you maintain an appropriate mineral balance.[21] You can find this as a supplement.

By addressing the lymph, the liver, and the kidneys, you'll have a good formula for an all-purpose detox, which is especially important if you are carrying a lot of physical or psychological stress: drink a healthy amount of high-quality water, take electrolytes or high-quality sea salt with it, supplement with magnesium and possibly some glutathione, and use motion or massage to keep your lymphatic system cranking. Now you have a system that can recover more quickly from whatever is wearing you down.

Hack: Cleanse Your Gut

I want to give an honorable mention here to one of the simplest but most effective detox hacks: activated charcoal. Activated charcoal doesn't help any one particular system in your body work better. What it does is stick like mad to a variety of toxins in your body before they hit your liver, your kidneys, or your lymph. It binds them in the gut, where they are expelled from the body, as you might delicately say. You can become more tolerant of toxins in your environment when you have activated charcoal in your body binding to those unwanted compounds. Because charcoal can bind a variety of substances and prevent their absorption, be sure to take it at least two hours away from other supplements, vitamins, and minerals and ask your doctor before taking activated charcoal if you are taking any prescription medications.

An important, related detox hack is using targeted compounds to bind and excrete specific toxins. I put humic acid and fulvic acid[22] into my Danger Coffee, both to pull coffee toxins out and to remove other toxins from your body. These are really interesting compounds.[23] They come from garbage—literally, humic acid comes from humus, rotting dirt. The molecules are small enough that they can pass into your cells (especially those of fulvic acid, which is a subunit of humic acid), where they can bind to metals so that your body is able to excrete them. They pull toxic metals and mold toxins out of your gut and your body. As a bonus, fulvic acid also acts as a nutrient transporter, enabling healthy minerals to get into your cells. I get humic and fulvic acids by drinking Danger Coffee daily. You can also take fulvic and humic acid liquid supplements that are easy to mix into water or tea.

Hack: Clean Up Your Environment

Your detox efforts won't be very effective if you live in an environment full of external toxins. I learned that the hard way when I was unwittingly living in and renovating a house built in 1908 that was full of mold. I was also dealing with emotional toxins from a disintegrating relationship, as well as stress toxins from my career and my mortgage payments. If I had been more self-aware back then, I would have removed more of those toxic elements. Instead, I carried so much stress that my body never came down from it to recover. My MeatOS never had a chance.

Most of the time, you have only limited control over your life situation. You can't just order your relationship to be good; you can't just insist that your employer pay you a lot more. You can't completely avoid volatile organic compounds (VOCs) and smog if you live in a city. But you can do things to manage your stress, to steer away from bad life decisions, and to identify and remove the environmental toxins you do have control over.

Most people don't realize how many toxins they are exposed to on a daily basis. Some of the largest yet controllable sources are household cleaning products, cosmetics, and body care products. Household and personal care products often contain large amounts of endocrine-disrupting chemicals that throw your hormones out of balance. By switching out toxic household products with natural, nontoxic alternatives, you can take a huge load off of your liver and kidneys.

As mentioned earlier, if you're living in a moldy environment, your liver and kidneys will be working in overdrive, attempting to process and remove mycotoxins that can damage your mitochondria and your brain. If you suspect you're living in a moldy environment, get your living space tested by a qualified mold inspector.

Chronic psychological stress isn't a chemical toxin, but it can prevent your detox systems from functioning at their best. If you're dealing with a large amount of life stress, adopting a daily meditation and breath work practice can increase your resiliency so you're better able to handle the toxins that you don't have control over.

After you remove the external toxins that you do have control over, you can detox in physical ways to give you back your energy and bring your baseline calm down to a lower, better level. Then you will have room to work the slope of the curve, taking the sharp, difficult actions that can improve your entire life.

Back in the day, I tried to do what a lot of other overachieving people do: I told myself I was just not trying hard enough. In reality, the harder I tried, the more I built up toxins and the further I got from the recovery I needed. It's one of life's wonderful ironies: as soon as you stop insisting that you can do more, that's when you start to get back the energy to *actually* do more. Slow down, take in some sunshine, get good sleep, meditate in a sauna, and the true improvements can begin.

ENDLESS IMPROVEMENT

11

SPIRITUAL STRENGTH

To be your best self—your optimal "normal"—you need to go after a sixth target, one that isn't covered by strength, cardiovascular fitness, energy, brain, or stress. You need to rebuild your *spirit* in tandem with every other aspect of your being. You will never be fully in charge of your MeatOS unless you improve yourself at this higher level. Physical resilience is intimately connected to the spiritual resilience necessary for you to practice gratitude, forgiveness, and kindness—essential elements of a peaceful and happy life.

I didn't always feel this way. I grew up in a very serious, science-minded family. My parents taught me that people are biological machines: meat robots, nothing more. Then, on a whim, I went on a ten-day personal development group retreat based on transpersonal psychology. I expected it all to be bullshit, but I was exploring. I was struck by one particular detail that one of the leaders of the group shared: many people on the retreat were also doing other types of cleansing in their lives, such as a heavy-metal detox. Many of them reported that they released a great deal of heavy metals if they revisited major traumas or painful relationship issues during the retreat.

That struck me as absurd. How could a week and a half of transpersonal psychology change the chemistry of the body and pull heavy metals out of someone's cells? That didn't sound like something a meat robot could do. But over the course of ten days of deep meditation, my thinking started to change. During that retreat I learned the technique of rapid, controlled holotropic breathing;

discovered how to relax; and was introduced to many different aspects of personal development. After I did that work, I slept better than I had in years. I noticed that I had far fewer cravings. It was easier to lose weight. I could be calm and relaxed about things that would have driven me nuts before, which was especially noticeable since I was going through a difficult breakup at the time.

All that I've done since then has led me away from the view that we are meat robots and toward an understanding that humans are actually creatures of distributed intelligence. The mind does not exist just in the brain but extends throughout all of our cells. We are held back by spiritual laziness as much as by biological laziness. No personal biohack will be complete without acknowledging that.

| COMING TO TERMS WITH EMOTION |

My path to spiritual healing began with a slow journey toward emotional healing. When I described my family as hard-core science types, I wasn't exaggerating. I come from a long line of engineers who were skeptical of anything you couldn't put on the table and measure. One memory that stands out: for her pleasure reading, my rationalist grandmother subscribed to the *Skeptical Inquirer*, a publication that showered disdain on anything that even hinted at mystery and unproven ideas.

Despite all that, I did some personal development work when I ran out of conventional things that could help me with my health and cognitive issues. After a while I had an epiphany: Wow, the mind is a whole lot more complex than I'd thought it would be. First I thought I simply had a willpower problem. I could fix that. Then I realized that my brain was broken: Oh, it's a hardware problem. Again, I could fix that. I was on the right track. If your biology is broken, your brain won't have enough power. If you fix your biology, you can start to fix your brain. But the physical brain is not the whole story.

As I kept doing personal development work, I began to see another level of complexity. Not only was I having emotional experiences, I was having spiritual experiences that sometimes I couldn't explain until years later. I started to recognize emotions as essential building blocks of who I am, not as garbage getting into my way.

Early in my spiritual journey, I did my first personal development workshop at the STAR Foundation, a very touchy-feely place. I was going through a painful divorce, and a friend of mine said, "Dave, you have to go do this." I asked what "this" was, and she said, "I'm not going to tell you, because then you won't go. Just trust me." I had been denying that emotions had any power over me, but I was falling apart. I was at my wits' end. So I took ten days off, which was a pretty extreme thing to do in a high-pressure Silicon Valley job, and left to do the workshop.

During one of the exercises that made absolutely no sense to me, the participants were doing primal therapy. They would allow themselves to go into a rage, grab a Wiffle ball bat, and beat up a pillow. I sat in the room thinking that that was the stupidest thing I'd ever seen. There were grown men and women, acting like unhinged children or animals. But I heard them making deep wailing sounds, letting go of all their negative emotions, and I could not deny that something truly powerful was happening. It turns out that primal therapy is extremely effective for some people. I still think it's not my thing, though. At any rate, I couldn't handle all the moaning and groaning, and I left the room to get away from the sound.

The head of the facility saw me and asked, "Why aren't you in the room? You don't have to participate, but you should at least sit there." I replied, "I just don't feel like being in there. I don't know why." She asked me to explain the feeling, and I couldn't, really. I told her that the exercise was stupid and I found it annoying. She persisted: "Is there a feeling in your body?" Yes, I admitted, my stomach felt weird. Then she said, "There's a name for that feeling. It's fear." That seemed ridiculous. I told her that I didn't have any reason to be afraid, so I couldn't be feeling fear. She laughed and looked right into me, a great spiritual master. "Fear is an emotion," she explained. "It doesn't have to have a reason."

That moment, that insight, opened up a whole new part of my

biohacking journey. I realized that people can be rational and emotional at the same time. The two can coexist, and the emotional side is there even when we try to pretend it isn't. So much of the time when we think we're being purely rational creatures, what we are actually doing is responding to a feeling and then making up a rational-sounding story to justify it. Just as your MeatOS directs your body to do something (such as pulling your hand away from a hot stove) a third of a second before your conscious brain can take credit (good thing I noticed and pulled my hand away quickly), so your emotions often respond first, with your reasoning racing to catch up. Either way, you need to know and respect the workings of your operating system and make sure you are hacking the cause, not the effect.

After the workshop and my early experiments with EEG, I realized that a big part of my biohacking explorations would involve mapping out connections between physical sensations and emotions. That was an important foundation of my work in spiritual recovery. Oftentimes, therapists and spiritual leaders say that we store trauma in our body: "Issues are in the tissues." This is a central message of Stephen Porges's polyvagal theory.[1] Trauma affects the long-term operation of the nervous system, especially the balance between the sympathetic and parasympathetic systems. Everyone is different, but many of us store trauma in our lower back and hips. Massage therapists report that women often start crying when their hips are massaged and cannot give a reason why. There is a reason, of course. (There is always a reason.)

Trauma is stored, unprocessed emotion. At the most fundamental level, it is a protective pattern your operating system puts into place to protect you from overwhelming experiences. Like your biological laziness, it's there to serve you. Unfortunately, the pattern-matching system in our brain is fast but stupid. It responds rapidly to sensory inputs, fast enough to keep us alive, but it doesn't understand the meaning of our stored emotions. Our conscious mind, on the other hand, is very slow but very smart. We are always playing catch-up with our emotions and reactions, wondering, "God, what just happened?" The stupid, fast system does what it does, and then we take credit for it, for better or for worse. Too often, we end up thinking we're bad people based on some stupid, self-preserving emotional response. The widespread concept of original sin may have emerged from the mismatch between our unconscious and conscious selves.

The emotional side of our biology presents no shortage of hacking targets. People trained in cognitive behavioral therapy help their patients learn how to engage with their feelings and emotions differently: when I get a feeling, I catch it, think about it, and decide what to do. That is an adult behavior, but it's a cognitively expensive adult behavior—all that thinking is exhausting. Then there's the Buddhist approach: I'm going to notice my feelings but let them go. I'll just observe and dismiss them. That's peaceful, but to my mind it misses the beauty of emotional recovery.

Or there's the path of emotional healing, which I prefer and is what I pursue at 40 Years of Zen. It is a path of forgiveness rather than dismissal. In this approach, the goal is to turn off the bad pattern matches so that you don't make inappropriate connections between events and emotional responses. It respects both the rational and emotional responses but aims to keep them separate. You can still have emotions, but they're no longer part of a reactive, emotional response. You can feel love, with all its marvelous giddiness, but you don't have to feel superangry when a guy cuts you off in traffic, your waitress says something you didn't like, or all the other things that become triggers for you throughout the day and make you grit your teeth. If you can keep from doing that, you have just freed up a huge amount of your neurological energy that can be redirected toward recovery, resilience, and growth.

ENTERING THE SPIRITUAL LAYER

It's not easy to capture the spiritual state in words, and it's different for everyone. It can be the feeling of dissolving into the universe or hearing a religious figure or a sage or a voice speak to you. It might be tapping into a source of universal knowing or feeling a sense of connection with something greater than yourself.

While the experience of spirituality is highly individual—and, for some people, ascribed or codified in a sacred manner—you have to know what's coming next. Yes, I believe we can hack our way to higher states of being. If we can hack our MeatOS, why can't we train our spiritual side? Why can't there be slope-of-the-curve improvement for spiritual growth?

You might say, Dave, that's sacrilege. But even the Dalai Lama has long suggested that science can abet the spirit, going so far as to address an annual meeting of the Society for Neuroscience on the subject. Likewise, biohacking has always been about more than molecules and electrons. As a part of my goal to upgrade my entire being, I've done ayahuasca in South America with shamans. I've traveled to Nepal and Tibet to learn meditation with the masters. I've fasted for days in a cave used by indigenous people for spiritual pursuits. Part of biohacking is exploring, and that includes exploring as far out as I or you or anyone else is willing to go.

Before you embark on a journey of spiritual biohacking, you need to get your energy level up and deal with all of the foundational diet, exercise, and training hacks. As you start working on your emotions, you will probably start to experience some deeply spiritual moments. You don't even absolutely need to fix your energy and strength and metabolism before starting on this journey. It's possible to have a spiritual experience when your body is profoundly broken. It's just much, much harder that way. It's better to have your MeatOS in order before you transcend your identity as a meat robot.

For emotional growth and healing, the slope-of-the-curve concept is just as important as it is in physical exercise: push into spiritual energy states where your mind doesn't normally go, and your operating system will adapt to become stronger, faster, and more resilient. If you go to a traditional therapist's office, you might have a breakthrough; then again, you might end up running through your one-hour sessions with similar, circular conversations for twenty years. You might get some self-knowledge that takes the edge off, but without ever achieving permanent change. But if instead you go to a short, goal-oriented workshop like the one I attended, you can go really deep quickly. Modern tools such as breath work, EEG and neurofeedback, light training, and sound goggles can rapidly push you into different spiritual states.

For example, it was once believed that it was impossible to induce gamma waves—the fastest brain waves—but today we know that this elusive state can be accessed with training. Gamma waves cycle up to eighty times per second; they are associated with sharp focus and concentration. Of course, those who are most successful at creating gamma waves tend to already have a high degree of spiritual enlightenment and self-control, but they're leveraging technology similar to neurofeedback in their training. Not everyone can become a spiritual master, just as not everyone can become a world-class power lifter, but everyone can improve the same way, following the same principles.

THREE STEPS FOR SPIRITUAL RECOVERY

When I say "spiritual recovery," I do not mean to suggest that it is analogous to addiction recovery (although successful addiction treatments often do involve spirituality). Spiritual recovery is the antidote to being spiritually stressed. What does being spiritually stressed look like? Maybe the love of your life breaks up with you, maybe there's a terrible accident, or maybe you lose your job. Maybe someone comes in and steals your company out from under you, or maybe a really famous podcaster drags your name through the mud in order to make money. The point is, despite your belief in the inherent goodness of humanity, sometimes shit happens that rattles you to your core. Your belief in Gaia, God, goddesses, the essential sanctity of life—whatever your spiritual framework—will be shaken. In order to recover, you need to detox and rebuild your spiritual strength.

If you don't recover properly from spiritual stress, you will operate from a place of weakness and confusion, alienated from your best self.[2] Even someone who is emotionally well put together, physically well put together, and mentally well put together can have a debilitating spiritual crisis. It can strike anytime you're filled with self-doubt: when you move away from home, when you get your first real job, or when you are about to get married, for instance. It can also be triggered by a wrenching loss. Many people have a spiritual crisis when a parent dies. Every parent I know who's lost a child has had a profound spiritual crisis. Spiritual stress will produce physical as well as emotional responses: rashes, pains, chronic fatigue, all sorts of crushing emotional effects.

How can you recover? You can't turn to the deer running from a predator for inspiration—except that you know what? you actually can. You just have to apply the slope-of-the-curve approach (fast signal, fast response) with the tools for confronting spiritual issues and embracing gratitude, forgiveness, and kindness. There's a specific three-step process that we have developed and tested at my my neuroscience-based brain upgrade facility. If you're open to hacking your spirituality, I can tell you that this protocol is incredibly powerful and worth exploring.

Step 1: Create Automatic Kindness and Forgiveness

We all recognize kindness as a virtue, but that is a vague and undirected value judgement. It's easy to decide whether you acted with kindness after you do something, but what makes you kind when you are acting automatically, from your MeatOS? Most people spend their lives trying to consciously think about every feeling they have before they respond to it. It is emotionally and physically taxing to do that; it requires a lot of work and a lot of electricity, and the whole role of emotions is to get you out of danger before you can think.

Wouldn't it be easier if you could edit out the parts of your OS that automatically make you reactive and mean? Instead of feeling, thinking, and then acting, you'd be able to just think or just act without having an inappropriate reaction.

Your MeatOS automatically defaults to behaviors that it thinks will keep you safe. A lot of the time, that default mode steers you toward selfishness, meanness, and unkindness. We've all seen human greed and war and the bad stuff that we do. The dark side of our nature is mostly your cells blindly running their operating system, guided by the three F-words. So what you want to do is to trick your spiritual laziness principle and reprogram yourself to automatically react with kindness. That is the spiritual path. That's the path that will lead to the smallest amount of suffering in your life.

To create automated kindness, you have to learn to forgive—and do it for real, not just as a set of words. Most people have an instant, negative emotional response to the word *forgive*. Whether they admit it or not, they think "I'm not going to forgive, because if I forgive that person, I might get taken advantage of again. I might get hurt again." Also, how can you forgive someone who did something really terrible? Would you forgive a mass murderer?

Using the true definition of forgiveness from spirituality, of course you would. That is what Jesus and so many other spiritual leaders taught: "Forgive them, for they know not what they do." So many people repeat those words without accepting the difficult work that comes with them. Once you place yourself into a mindset of true forgiveness, you could look at a mass murderer and say, "I wonder what his dad was like. I wonder what it's like to be that person and what terrible things have to be happening in his head and heart to make him act like that. I wonder how much he's suffering. I'm glad I'm not him."

Forgiveness is *not* condoning something bad. It is also *not* telling a person that what he or she did is okay or even that you forgive him or her. Forgiveness is simply letting go of a grudge so that it no longer causes you to lead with reactivity. That's it!

When you truly forgive, your entire state shifts. That change is something that you can feel in your chest and is observable in your brain wave patterns. When you forgive, you permanently remove the automated unkindness response from your operating system. It's like going into your phone and turning off alerts. That will be a massive upgrade for you. Try turning on alerts for every app on your phone, and see how sane you are after three days. You can't think straight when every email, text, social media post, app update, and so on keeps going boing, boing, boing. Your MeatOS is doing that to you right now, and you won't believe how peaceful your life will become when you learn to silence it.

Step 2: Activate a Spiritual Reset

When you set out to learn true forgiveness and create automated kindness, the spark that lights the fire is gratitude. You could say that gratitude is the energy that can push you up the slope of the curve to forgiveness so that you can achieve a spiritual reset.

Let's say you get into a terrible car accident. You might say, "All right, I'm going to drop my trauma. I'm going to drop my fear response to cars and driving," which is an anxiety that many people experience after being in an auto accident. But letting go is only one part of the process. You also need something positive to hold on to. You could say, "I'm glad that I'm alive. I'm glad that no one died." You need to find one good aspect of the situation. It can be something small; that doesn't matter. It just has to be one thing, because as soon as you give your brain gratitude, your cell biology will shift to a state of receptiveness.

Once that happens, you can run through the spiritual reset process. This is a physical biohack in addition to a mental one. What you do is sit down, close your eyes, and take a few deep breaths. Get yourself into a meditative state; you may find it helpful to do this with a guide, but it is certainly something you can do on your own. Then you seek out a memory, a mental wound, that still triggers you. Pick the time you were bullied in fifth grade, any incident that still feels painful. If you don't know where to start, that's fine: pick the very first thing that pops into your mind, no matter how big or small. We all carry thousands of little triggers, many of which seem absurdly small when you try to describe them. When you do this work, you will find yourself remembering the weirdest things. Those memories are your mind and body telling you what you need to tend to next.

Now envision yourself the way you were when the event happened. If it happened when you were ten, imagine yourself as a ten-year-old. If it happened yesterday, you can envision yourself as you are now. Envision the other person from the memory sitting across from you—if it's a person. Usually it is, but it could be a group of people.

Say out loud (even if you are by yourself; in fact, it works especially well when you are alone), "You did this, and it caused me this harm." Allow yourself to feel bad. The discomfort and pain have to be there. You can't just imagine them rationally. You've got to really connect with the hurt.

Then find the gratitude. It was awful, but one good thing did come out of it. There's always one good thing. Once that happens, put yourself into the other person's shoes and ask yourself, what the occasion was like for them. Many of us have forgiveness work to do related to family events in childhood. Those memories can require a radical rethinking of the other person's perspective. When you are three or four years old, sometimes you get pissed off for no good reason. Your parents probably couldn't even tell. They were just trying to buy groceries, and you lost your mind and now it's stuck. Now get unstuck. Look: There's your mom. She was really hurried, she was doing her best, she had no idea. You never thought about her experience because you got the emotional setting when you were a small child. Now you are thirty, and you can make her experience your own.

Next, look at it from the other person's perspective and your perspective at the same time. At this point, most people focus on sending a light beam or some other kind of signal, emerging from their chest area and opening up until they really can see things from the other person's point of view as well as your own. When the connection feels wide open, so that you can see things from the other person's point of view, you say, "I forgive you." You will feel a relaxation in your chest; it's extraordinary. When we run these sessions at 40 Years of Zen, we register a marked change in brain waves at this stage. You can see forgiveness happening in the body.

Step 3: Advance from Forgiveness to Acceptance

For the final step, you need a neutral third party, a judge, a referee: someone infallible, therefore not an actual person. You can make it Jesus or Buddha. Hell, you can make an infallible light bulb if you want.

When you think you're done forgiving someone, with your eyes still closed, go to your judge and ask whether the forgiveness is complete. If you get a "yes," you're done. And if you get a "maybe" or an "eh," you go back and keep doing it. This is where most people fail. They say that they forgave, but they are still crossing their arms in front of them, still angry. Forgiveness is not cognitive. It is spiritual and emotional. Simply saying that you forgive isn't enough. When you truly forgive, though, it unlocks the emotion that's causing spiritual stress.

What you will end up with is complete nonreactivity to something that would've been triggering for you before. That change will enable you to have more energy and more freedom to perceive and experience spiritual states. Everybody's response is different—and I say this having looked at 1,200 high-resolution brain scans—but everyone has the capacity to become more enlightened.

An essential point to understand is that spiritual rejuvenation is different from physical rejuvenation in one extremely significant way. With physical enhancement, you can pursue your target directly. If you want to be stronger, you set out to become stronger, you do the right hacks, and you become stronger. It's straightforward. With spiritual enhancement, if you come in saying "I am about to make myself spiritually strong, I'll be a goddamn forgiveness powerhouse," you are on a fool's errand. If your motivation is to make the world a better place, on the other hand, the benefits may just sneak up on you. I remember doing an intense forgiveness meditation once, and when it was over, I looked around and felt as though my intuition level was ten times higher than before.

These kinds of breakthroughs are always centered around forgiveness. The more alerts you turn off, the more you can pay attention to your connection with the earth and with other people.

Kindness becomes automatic. You give more to the world, and, amazingly, it takes far less energy to do it.

Gratitude, forgiveness, and kindness form a triad. Instead of having to choose kindness, it happens automatically. You can look at a homeless person and think "That is a bad person, why doesn't he just get a job?" Or you can think "That guy looks thirsty, I'll give him my bottled water." If you record a video of yourself doing it and post it on YouTube so people can see what a good person you are, you really weren't that kind. But with automated kindness, you can be generous just because it was easy and it was your first instinct. That doesn't make you a sucker. It makes you strong.

Once you have completed the third step of the reset process, other people and stressful situations are much less likely to trigger you. The result is that you will have more freedom. You will be kinder but also more dangerous. People with fewer filters, people who are automatically kind, are better at reading the personalities and situations around them. They are more resilient because they respond to what is real, not to old triggers stuck inside them. A great side effect of spiritual recovery, therefore, is that it gives you an extremely effective bullshit detector. You can smell propaganda a mile away. When someone is trying to manipulate you, somehow you just know.

| SPIRITUAL GROWTH HACKS |

One of the most interesting words in the English language is *ineffable*. It means that there aren't words for something, so we actually have a word for not having words. Most spiritual states are ineffable. For more than a thousand years, people tried to make the ineffable effable with specific chants, visualizations, and the like. There are thirteenth-century Sanskrit writings that provide precise instructions on how to induce these spiritual states. Today we can use technology to help induce such states, which is a profound acceleration in spiritual growth. Outside of being hooked up to a fancy machine, though, there are a number of ways to enter a spiritual state—and the more you do it, the easier it becomes. Here are a few of them.

BIOHACK SUMMARY

- **Spiritual guidance** by a guru can be valuable as long as you don't attach yourself to one teacher for your entire journey.

- **Breath work** such as holotropic breathing alters your state of consciousness to unleash intense emotional and spiritual experiences.

- **Having great sex** or even experiencing a full-body orgasm can be a spiritual experience.

- **EMDR**, or eye movement desensitization and reprocessing, involves moving the eyes back and forth to process and integrate past trauma.

- **Psychedelics** come with biological and spiritual risks, so if you are going to do them, be sure you do them in a spiritual setting with the right experts around you.

- **EEG**, or electroencephalogram, enables you to see the patterns of trauma in your brain waves and can help you effectively train your brain out of those patterns.

Hack: Spiritual Guidance

Should you follow the old paths and work with a guru? I don't, but many people do. I learn from as many different lineages as I can because I'm a biohacker, I'm an explorer. I've gained great value from different teachers at different times. I think that latching yourself to only one teacher comes with risks; teachers who demand exclusive loyalty may have a lot to offer, but there's also a downside to that versus the ones who are happy to share whatever. When you're there, you're there.

Hack: Breath Work

Breathing exercises are helpful for many of the biological hacks, and it turns out that they are important for spiritual hacking as well—especially holotropic breathing or Wim Hof breathing, the aggressive, deep, fast breathing that is so powerful in combating other forms of stress and enhancing neuro work (see page 164). I first experienced holotropic breathing at the personal development event at the STAR Foundation. I left my body. I had profound visions that I wrote down, including past lives. You can assign great value and meaning to these things, or you can say it was just random electrical firings. But for many people, these simple-seeming breathing exercises unleash intense emotional and spiritual experiences.

Hack: Having Great Sex

It's worth noting that for many people, sex can be a profoundly spiritual practice. About 20 percent of people report meeting God or leaving their body during really good sex. There are plenty of biohacks that can help you have better sex. Pretty much every piece of advice in this book is relevant: greater strength, more energy, a stronger nervous system, and less stress are all prime ingredients for a more rewarding sex life.

Some people ask whether there are ways to attain the spiritual pinnacle of the orgasm without actually having sex. Is there a way to achieve an expanded state outside your body? That would be a very powerful thing to do, and some people are doing it. It turns out that you can achieve a full-body energy orgasm or a spiritual orgasm by directing the energy flow in your body inward and upward. You can learn more by searching online for tantra yoga or tantric sex.

Hack: EMDR

Francine Shapiro, a PhD psychology student, was watching a tennis match when she stopped moving her head, started moving her eyes, and went into a startling altered state. She immediately changed the topic of her PhD thesis and discovered a previously unknown reset mode in the mind. It is called eye movement desensitization and reprocessing, or EMDR.

Today, there are thousands of EMDR therapists all over the world. You can find one near you on the EMDR International Association website. The technique superficially resembles hypnosis, but it works very differently. A therapist will instruct you to move your eyes back and forth, usually by having you follow the movement of their fingers, and sometimes he or she will touch your knees or have you hold little buzzers that make your attention toggle back and forth between the two sides of your brain. While you do these things, you think about a situation that feels extremely triggering to you. The movement and sound treatments call up negative emotions and desensitize you to the trigger. EMDR helps many people process and integrate intense traumas. I have tried this therapy myself and found it to be dramatically effective.

Hack: Psychedelics

One of the riskiest ways to induce a spiritual state is to use plant medicines. But if you decide you want to experiment—and if it's legal and available in your part of the world—you could start out with MDMA. After that, in order of increasing risk, you could look at ketamine, GHB, and then mushrooms, LSD, DMT, and then ayahuasca. In addition to their biological risks, all of them pose a risk of spiritual harm, according to pretty much every teacher I've ever worked with, even the shamans who use them in their practice. My advice is that you don't have to go there. But if you decide you really want to go there, I urge you to use plant medicines in a spiritual setting with appropriate spiritual people around you. In other words, don't do them at Disneyland, and absolutely do not do them alone.

Tech Hack: EEG

With an electroencephalogram, or EEG, you can see the patterns of trauma and PTSD in someone's brain waves. Then you can train the brain out of that pattern. You may recall that I mentioned EEGs earlier, in the context of neurofeedback. Promoting brain health and promoting spiritual wellness are related, but they are not the same. To most people, brain strength means thinking quickly and clearly, remembering details, being able to perform complex work or creative tasks. Those things are all important, but they miss the underlying issues of pain and anger that hold us back, as well as the kindness and equanimity that allow us to embody our very best qualities.

We all carry mental wounds that we're not conscious of. You may be afraid of something. You may be mad about something. You don't know that the feeling exists because someone almost dropped you when you were two years old. It's outside your conscious awareness. EEG in combination with personal development work is a powerful way to heal from trauma because it allows you to examine the trauma the same way you would examine a brain wave pattern that prevents you from moving your leg fully. You get to choose the state you are in, based on the EEG readout. Unlike with EMDR, you are not in reset mode; you are in cancellation mode.

It's possible to create a state that cannot coexist with feeling traumatized. I might say, "I'm back in my childhood, looking at Mommy yell at me, but I'm also in an exalted spiritual state because the computer put me there." My body feels the spiritual state and decides that I should just be there—like pushing your muscles past the point where your proprioceptors thought they should go. All of a sudden, things that previously upset you don't seem so bad anymore. I know someone who did EEG treatment before spending the holidays with his family. He called me in tears and told me, "I just had Christmas at my parents' house, and for the first time ever, I didn't get in a fight. I didn't think it was possible. How the hell did EEG do that?" His brain had been stuck in a reactive pattern until he learned how to control his thinking and to escape the trap.

EEG can reset automatic invisible patterns in the way you interact with the world. This is another way that biohacking can give you back your lost time and money. Let's say you have a hundred neurological energy points to spend. You could spend half of them reacting to stuff that isn't real. That's what most of us do a lot of the time. You might think "That guy cut me off in traffic because he's a bad person and I want revenge." Or you might think "That guy cut me off in traffic because he is on the way to the hospital and is frantic." You don't know which one is true. But if you believe the first story, you blow through your neurological points by playing out a stupid encounter over and over. Those are points that you could be using to create something beautiful or love somebody or perform a positive service.

Trauma is a drag on your neurological recovery because it's always present and it gets triggered easily, sucking up energy that should go into *being you*. Instead, it's going into playing an old broken record that isn't even accurate anymore.

Although your spiritual state is an intangible thing, it greatly influences the physical state of you and your MeatOS, which means that you can measure it and evaluate it. The best data point to look at is your heart rate variability, which you can monitor using a simple tracking device. If you see improvements in your heart rate variability, you are probably moving in the right spiritual direction, because generally the more spiritual you are, the less time you spend in fight-or-flight mode; you will have higher heart rate variability, because you will have more built-in adaptability. Improvements in deep sleep, improvements in REM sleep, and remembering more dreams are also good indicators.

When you raise your spirituality, you'll feel more connected to the universe. When this happens, you likely won't be relying on fancy tech or physical data for confirmation. You will just know.

THE NEXT-LEVEL UPGRADE

A core part of the biohacking process is continuous improvement. You want to keep refining and upgrading yourself through the cycle of evaluate, personalize, repeat. The same is true of biohacking itself. We biohackers are always looking for ways to go deeper into the MeatOS by applying new science, new techniques, and new technology. We are still learning how to extract more of the energy, resilience, and clarity that are locked away inside you—to find even better ways of co-opting the laziness principle—to go beyond anything that's possible with biohacking today. Here's a sneak peek at what I see coming next.

A QUANTIFIED SELF

One of the most exciting trends in biohacking is the ability to gather and process tremendous quantities of data about our individual physical states, known as the *quantified self*. By gathering enough data, we can validate human experiences that scientists long dismissed. EEG devices and other measures of brain activity, for instance, show that the spiritual states attained in Eastern meditation are real. Even though everyone doesn't sense them the same way, we can measure them. In addition to electrical change in the brain, we can register physiological changes and changes in people's blood when they enter these states. That level of detailed measurement has the potential to transform our understanding of what humans really are. And when we have more understanding, we have more opportunities for control.

We also now have the data to show which things probably do not work. Traditional forms of exercise—the endless gym workouts, spinning classes, and other entrenched practices that I call collectively Big Exercise—are being exposed as wasteful and unproductive. Over the past decade, medical researchers have started to develop a completely new idea about how fat works in the body. It turns out that it is not so bad after all—at least, not all forms of fat. Many recent studies are revealing that brown and beige fat, which are loaded with mitochondria, contribute healthy heat, energy, and building block compounds to the body.

Millions of people are now contributing to our knowledge about human health and fitness. If you wear a fitness tracker or use a blood glucose monitor, you are probably contributing. If you have a health-monitoring app on your phone, you are almost certainly contributing to a global database, whether you know it or not.[1] Never before have we been so aware of the collective wellness of our species, and we are just getting started.

Every day, health-monitoring technology gets cheaper and cheaper. Soon your doctor will be able to map everything that is living in your gut, from bacteria to viruses to phages to fungi. Then you could be prescribed prebiotics and probiotics designed to enhance your microbiome and pump up your metabolism. For example, if you eat a lot of pomegranate and you happen to have the right bacteria in your gut, they will manufacture a powerful antiaging compound called urolithin A. But you have to eat pomegranate, and if you don't have the right bacteria (which most of us don't), all you get is a lot of sugar. If you had a complete map of your microbiome, you could get a dose of whatever good bacteria you are missing, and then you could take personalized supplements to feed them and make them work optimally for you.

Going forward, medical sensors will be everywhere. You might play a video game that can track your eye movements and diagnose that you've been exposed to a neurotoxin or have very early indications of Alzheimer's disease. Today's simple blood glucose sensors will become much more powerful. You will have a whole lab on a chip that you can wear on your arm, monitoring your biochemical status from moment to moment. We will all know more about the state of our biology and therefore will have more control over it than ever before in human history.

SELF-TRACKING

The coming flood of medical data will lead to sweeping changes in how we can track ourselves. You are a member of a vast collective of *Homo sapiens*, a complex, diverse species. When we combine self-tracking data from millions of people, we will see unimaginable new patterns. Any ER doctor or police officer will tell you that the full moon really has an effect on people, even if only because it changes the lighting and mood outdoors. There's a huge increase in the number of people who go commit crimes during the full moon because it does something to us. We've known that forever, but scientists generally haven't taken the stories seriously because they can't quantify them in a lab study.

With self-tracking data, anecdotes will soon become established fact. Who knows what else we will discover? When we watch a million people's biological readings shifting all the time, we will be measuring not just one person's change but a change in the human condition all at once. Then we can use machine learning systems to figure out which things are most highly correlated and look for causative factors. We're going to find all sorts of stuff that's affecting us that we don't know is affecting us right now. That's ultimately where self-tracking is headed.

PERSONALIZED INTERVENTIONS

Quantified-self technology combined with large-scale data tracking allows for extreme personalized interventions. When you go to the doctor or a fitness trainer, that professional will have access to information from people all over the planet. He or she will be able to filter hundreds of millions of people's data points to know what will work best specifically for people like you. That's why self-tracking is so powerfully important. It doesn't just tell you your state; it tells the health professionals what caused your state, and they can use that knowledge to help other people.

In the future, an arm-mounted lab on a chip could measure your biological aging process and recommend antiaging compounds that are effective for people with your type of genetic makeup and environmental exposures. It could also serve as a great preventive tool. Alternately, your future health monitor might be woven into your clothes.[2] Constant monitoring means that we would be able to recognize the very early signs of Alzheimer's or heart disease and intervene quickly before it becomes an issue.

The Dangers of Testing and Data Collection

There is a dark side to the rapid expansion of testing and data collection: your information could be misused by companies and governments. If we go down the dark path instead of using all of this knowledge and technology to give people control of themselves in the pursuit of happiness, we could end up giving control to someone else. It's a scary possibility, one that would remove happiness for everyone.

As hackers, we know that the way to make sure we don't go down the dark path is to know what is possible, so that we can use it to our advantage. I'm optimistic about emerging technologies such as the distributed ledger, a shared and distributed database (related to blockchain technology) that does not belong to any one user or institution. In a distributed ledger, when knowledge or information is put out into the world, it cannot be censored or deleted because it's in too many places at one time. If you remove part of the record, the next part of the record can't be read. That way, if you were to discover something important about the human condition with your data, your finding would be distributed across everyone's devices all over the world. No one could take it away or hide it.

What if we fail? I hope it doesn't happen, but the world of personal data could go very wrong. Imagine a world in which companies or governments can keep essential medical knowledge secret. Here's an example. Back in the 1980s, the Russians figured out that humans are susceptible to certain wavelengths of microwave energy. They bombarded the US Embassy in Moscow with tuned microwave radiation so that the employees there would get foggy and sick. The embassy had to rotate personnel in and out every three months. A team of US specialists finally figured out what the Russians were doing and ordered them to stop. The Russians smiled and replied that their microwave transmissions fell below the allowable health standards in the United States. They refused to stop, because they were doing something that was perfectly legal in the United States. The Russians knew that microwave radiation affects us biologically. We didn't, and we suffered as a result.

What other technologies are out there that can change your mental, physical, or emotional state? There may be ones that we are unaware of because we don't know they're possible. We have to know, all of us, openly. With ubiquitous tracking data, it will be possible to know that something happened and look for the cause. You could look into the databases of the fitness-tracking companies, of people doing very basic EEG readings, of people measuring their heart rate variability, and so on. You could compare today's data to historical trends, looking at average increases or declines in people's overall readings. You could notice if something strange happened. For example, if yesterday the heart rate variability score of the population dropped by 8 percent from the average, something big must have happened.

Then we can unleash machine learning algorithms to sort out the cause of the anomaly. The big question is: Who will be in charge of the machine learning, and who will be in charge of sharing and interpreting output? Will it be a government? Will it be a tech monolith? Or is this part of our basic human utilities that should be open and available to the public? I strongly believe that in order to be happy, you must have control of your own biology and be free to choose what you subject yourself to. Open data sources are vitally important for maintaining and improving human consciousness.

| EPIGENETICS |

Epigenetics is an exciting, relatively new science that studies how environmental factors can switch your genes on or off. Genetics tells you what the switches do; epigenetics tells you what you can do to make them active or inactive. Epigenetic triggers can instigate inflammation, for instance, turn up your biological energy production, or activate the production of biological compounds that fight aging. Researchers are still struggling to figure out exactly which inputs affect the epigenetic switches and how they work. Data from fitness trackers and other medical sensors will make this work much easier.

One way to learn a lot about your epigenetic risks is to look at your exposome, a compendium of all the environmental factors that influence those switches. We're starting to do this with fitness trackers, with tracking mobile phones moving down the freeway, and with other ubiquitous devices. Air quality and weather monitoring help determine your exposome. Even space weather—magnetic storms from the sun blowing past the earth—seem to have an effect on how we feel. Do they affect our epigenetics? Nobody knows yet, but they will.

New insights into epigenetics are already changing the way medical research is done. Many, many medical studies are done on mice. The scientists claim that they account for all of the variables, but it turns out that they have been missing some truly huge ones. Mice are nocturnal creatures. Scientists are awake during the day, however, so that's when they run their tests on the animals. They have been feeding and experimenting on mice while, according to their internal biology, the mice should be asleep. That's an epigenetic factor that needs to be accounted for. The physiological response and stress levels of mice are also radically different if a female scientist rather than a male scientist feeds the mice.[3] That's another long-ignored epigenetic factor.

Decades of studies now appear suspect,[4] yet the results of those studies still affect the drugs we use and the ways we think about disease. These are invisible, epigenetic effects that most of us don't think about but that are affecting our reality.

| ARTIFICIAL INTELLIGENCE |

My undergraduate degree is in a subset of artificial intelligence (AI), and I've been following progress in the field closely as it takes on a growing role in helping us make decisions. These days, your phone recognizes your voice by using an embedded AI on a chip. It's amazing how far we've come and how rapidly the technology is still advancing.

There are two aspects of AI. One of these is machine learning, the data-sifting process I referred to earlier. By sorting through enormous data sets, AI can find correlations that no human would ever see or even think of. Humans naturally seek out a single variable that may be causing a symptom or a physical improvement. In reality, there are always multiple overlapping variables at work. Life is complicated that way. Machine learning can uncover those overlaps in ways that the human brain simply cannot. Until now, the main limitation with machine learning algorithms was that we did not have enough data to feed into them. Now we do.

When someone comes into Upgrade Labs, for example, we can suck in data from their health trackers. That gives us many thousands of data points from each visit, whether it's specific strengths or specific VO_2 max. What is the state of their inflammation? What about their fat storage? How well are their cells performing electrically today versus during their last visit? We can gather all of those data and put them into the system. People can get a full status report that shows up on their phone app; at the same time, they are contributing to an extremely useful sum of knowledge. Over time, the system will get better and better at shaving off time and effort from your biohacks.

If we are really smart about applying machine learning, we will eventually get close to 100 percent effectiveness. We will be able to tell you: If you apply this particular signal to your body at this particular frequency and duration and power level, you will get the fastest possible biological results. Then we can explore the effect of additional commands to your MeatOS: a peptide, a performance-enhancing drug, a different form of radiation. We still don't really know the upper limit of how much you can improve yourself.

Here's a far-out thought: I would like to see a new "upgraded league" in sports. Let's take the gloves off and see what humans can do if you allow them to compete without limit, using biological modifications, performance enhancers, anything. The only rules would be that they would have to tell us what they did. Rather than holding ourselves back with the Luddite view that players should be pure in some sense, let's really see what we can do as a species. It's not as though the current system is fair anyway, with younger players pitted against older ones who have lower hormone levels and more mechanical wear and tear. I think that we're going to see an increasing demand for upgraded players. Soon almost everyone will be upgraded in some way. If professional athletes are the only exception, what exactly is the point?

BIONICS

Speaking of enhancement: there is a subset of biohackers called grinders who buy or manufacture devices that they implant in themselves. Journalists love them (humans becoming robots!), but I shudder when they are included in the world of biohacking. So much of what they are doing strikes me as dangerous grandstanding. The reality is that a lot of people are already bionic. I have a screw in my knee and a screw in my foot. Many people have brain implants to control seizures or severe depression. Almost everyone has the world's number one brain extension, the smartphone. We no longer store phone numbers and map directions in our brains; we have distributed that part of our memories into our phones. Every search engine on the planet is an extension of your brain, no implant required.

I consider implants dangerous because any active, networked device in your body can and will be hacked. We've already seen instances of people hacking control systems for cardiac implants that didn't have any security provisions.[5] Or what happens when the company that made your cognitive implant is acquired by Google and Google decides to shut the division down? What are you going to do, pull it out? We've already seen this happening. Some people who had received retinal implants couldn't see anymore because Second Sight, the company that made the implants, went out of business and could no longer support the tech.[6] It's worse than not having had an implant in the first place. Just imagine the day when spammers put malware into your bionic eyes so that you see their ads everywhere or a judge decides that your implant shouldn't be able to see certain things.

There are also major health risks to having an implant. One is biofilms, aggressive bacterial infections that form a collective defensive shield so that they are very hard to kill. Biofilms are already a major issue for medical implants. A second health concern relates to the materials in the implants, which can trigger reactions in the body. Some people's immune systems go crazy in the presence of certain metals or plastics. The response is not very predictable. Adding to the problem, manufacturers are not always good at disclosing the composition of their implants. "Titanium" implants can contain up to 5 percent nickel. If you have a nickel allergy, that is enough to cause serious harm. Researchers still haven't solved the problem of rejection of things as basic as bone screws and breast implants.

My third health concern is the effects of electromagnetic fields (EMFs) generated by the implants. EMFs can affect the tissues in your body, your brain especially, through a mechanism called the voltage-gated calcium channel. We don't know all the ways that various EMF frequencies affect your cells. Having a device inside you that's constantly emitting EMFs seems like a bad idea. It could produce side effects that will take years to detect.

For people who are blind or deaf, an implant may be worth the risk. For people who are just looking for augmentation, a much more promising solution is a technology I experienced at Abundance360, the inventor Peter Diamandis's group for futurism. Last time I went, there was a group showing off a contact lens that you could hold up against your eye to give you night vision or perform facial recognition. It's not wearable yet, but the inventors are getting close. My recommendation is to hold off on biological implants as long as you can. Extreme wearables will give you most of the same benefits without the risks. Hopefully, by the time implants are common, we will have solved a lot of these safety and privacy issues.

| BRAIN READERS |

Another fascinating and creepy technology is Elon Musk's Neuralink, a brain implant his company is developing.[7] Given all the issues I just laid out, I really don't want microfilaments inside my brain. I would want to see a lot of safety studies before considering any kind of implant, but especially in the brain. Look back at the cautionary example of breast implants. People have been getting implants for forty-plus years, but only recently have the companies that make them openly acknowledged that their implants can cause disease and serious autoimmune problems. Did you want to be the first person to get breast implants, or did you want to look at a couple decades of data so you could make the decision with informed consent? The risks are vastly larger for anything going into your brain. I'm going to want to see a lot of other people be guinea pigs before I decide whether to do so.

External brain readers are a whole other matter. We can already do this based on blood flow, electricity, and other changes that we can pick up with sensitive instruments. Bryan Johnson, the CEO of Kernel, is developing one of these.[8] Bryan was one of the founders of

Braintree Venmo, a payment-processing company, and made enough money to put $80 million into building the best brain-scanning system ever developed. He has created a helmet that can track what's going on inside the brain, so that scientists can learn more about how we think and AI systems can be better trained.

Eventually you may own a brain scanner built into a hat or, even better, into the ceiling of your office. It will read your brain waves and give you a real-time assessment of your mental status. There is no technical reason why that can't be done. Having more information will help you take full advantage of the hardware that is already in your body. We owe it to ourselves to explore the profound untapped capabilities inside ourselves before we start ripping out parts of the body and replacing them without even having turned them on.

A DIGITAL VERSION OF YOU

I'm confident that our technology will eventually get to the point where you can take a snapshot of your brain activity and upload it to an AI system. But make no mistake, that AI will not be you! Your brain is distributed throughout your entire body. You are not your brain. You are your body and the way it interacts with your environment. That said, would I love to have an AI version of me that could think and collaborate with me? Hell, yeah, but I'd always remember that it is just a digital simulation of me that can help me get stuff done.

Maybe at some point, that AI version of me will be a ghostwriter working on my books. My AI counterpart won't need to take breaks and could work 24/7. Not only that, it could run in parallel on five hundred computer processors at the same time. I might able to say, "I want to write a book about topic X." Then the next morning, the book would be done, written in my own style. That would be cool. Of course, everyone else would be doing the same thing, so the world would be flooded with books. Then I would need my AI to go on a reading binge and tell me which ones are the most interesting.

I'm currently in the process of doing a simple version of this. I am uploading all of my podcasts, my writing, and my social posts to an AI system that will then comb millions of news sources to find stories that match the topics I'm interested in. Then I will turn that information into a newsletter. So basically, I'm taking a rudimentary digital Dave and curating profound amounts of information, filtered through my lens on reality.

You are already doing a similar kind of project without even knowing it. Every time you go to any social media page, it collects information about your likes and dislikes, your behavior patterns, the things you buy, the ideas you find interesting. When you fire up Facebook, Instagram, or TikTok, you've already outsourced a lens on reality, based on a bunch of links you clicked on and handed to a company that is trying to sell you things. I recommend opting out of that and using a search engine that searches the information sources themselves. In the future, upgrading your information environment will be an increasingly important reality hack.

THE NEXT ERA OF BIOHACKING

Despite these cautionary notes, the data explosion has mostly led to huge benefits in how we live and in learning more about how we could live better. For a decade now, I've been writing about the benefits of cold therapy and the best ways to do it, but there wasn't a huge amount of academic research on the topic. In the past couple years, though, a bunch of academic research has come online. Researchers are talking specifically about exactly how much heat, how much cold, and how frequently to apply it in order to extend human life. They are sharing ideas and posting their data publicly. As we gather and share more data about all aspects of human health, we are learning vastly more about how to hack the MeatOS to make it do what we want it to do.

We can now test word of mouth and ancient knowledge, and do it at a breathtaking pace. Researchers are confirming ideas that were dismissed as superstitions, and their AI systems are picking out correlations and effects that everybody missed. For example, doctors are already using AI-guided medical diagnostics to make more precise diagnoses of certain diseases such as cancer and to select the most effective treatments for each specific case. In the old style of exercise, you would go to the gym and the trainer would basically tell you to do whatever Arnold Schwarzenegger did thirty years ago. With innovations such as AI-driven exercise machines, we can push the slope of the curve and do a lot better: more benefit, less effort. But we still have so much to learn. As a trainer and a biohacker, I want to know the exact speed at which you should raise and lower your arm and whether you should turn it at a certain point, optimized for your arms, your metabolism, and whatever you ate last night. Then I want to be able to tell you exactly how much benefit you got from the exercises and how much time and effort you saved, because humans respond really well to evaluations and rewards.

From my glucose sensor, I learned that my blood sugar level goes up 20 points if I eat a half hour after the sun goes down versus a half hour before the sun goes down. That information has made me conscious of something that I felt impressionistically before but that I now know for certain via quantified, real-time feedback. I noticed that certain eating patterns caused me to gain a couple pounds and slowly nudged me toward being prediabetic. With hard data in hand, I changed my meal timing, all because of the quantified data that I had never had available before.

Another example: I used to take notes on all the things I did during the day that I thought might affect my sleep; then I would wait three months and draw a graph of my changing sleep quality, looking for correlations. I did that for fifteen years before I had enough information to get my sleep roughly optimized. Today, I continuously play with new supplements and supplement timing, taking things out and adding things back in, because I can get good data from my sleep tracker almost immediately. I added a PEMF device when I go to bed, because it triples my heart rate variability for the first hour or two of sleep. I'm experimenting with vibration wearables such as the Apollo watch, and soon I'll be able to buy light stimulus devices that will be tuned to flicker in a pattern that will produce clinically significant results for me. I used to dream about being able to hack my body using such precise inputs and such detailed readings about how my MeatOS is responding. That dream is coming true.

We're in unexplored territory about how far we can take our bodies. The first wrist tracker that could monitor your heart rate came out just a decade ago, and now Apple and other companies make tens of millions of watches that do the same thing. Now there are groups of inventors exploring ways to build wearable devices that apply pulsed magnetic fields to specific parts of the brain, no wires or implants required. Just imagine where we will be a decade from now.

I want you to embrace the ideas in this book because I truly believe they can make you stronger, kinder, and happier. They can give you the resilience and equanimity that will allow you to handle change—and believe me, change is coming. On its own, your lazy operating system may not be prepared for the next ten years. With the right hacks, though, you can be ready for anything.

YOU DO YOU

Diets come and go; that's why there are so many new diet books coming out all the time. Exercise fads come and go; that's why so many people join a gym, get fed up, quit, join another, and so on. But biohacking is not a list of tasks or rules. It is a process that belongs to you, so it inevitably grows and changes with you. Best of all, biohacking is based on measurable results, the most important of which is: It should make you happier.

This is such an important point. It brings us back to the reason why I became a biohacker in the first place, the reason why I've worked so hard to take control of the human operating system, and the reason (whether you know it or not) why you are reading this book. You want energy so you can be happy. And I want you to be happy. Maybe you're a spiritual master, maybe you're not. Maybe you're strong and want to be stronger, or maybe you can barely roll out of bed in the morning. Regardless of their state and situation, people just want to be happy—certainly, happier than they are now. What makes me happy is my family, my work, and biohacking. If what makes you happy is painting, then that is what you can get more of from this process. You want to be the world's best painter? Turn off your phone alerts and have more energy.

I have been dealing with trauma since birth, literally. When I was born, my umbilical cord was wrapped around my neck. Before I learned about spiritual recovery, people who study the effect of birth on psychology could see the trauma clinging to me from across

a room. Trauma leaves a visible imprint and creates a noticeable behavior pattern. It sets up a lens on reality that you are always in
danger. The laziness principle in your biology already primes you to
view the world as an endless succession of threats, but experiencing
trauma makes it worse. People with unresolved trauma walk around
with an anxiety so deeply rooted that it feels as though it's reality.

If you're more reactive and more threat seeking, you'll find more
threats. For me, being the largest, fattest, tallest kid in school, I always attracted the bullies. I got into a lot of fights. I never threw the
first punch. I almost always threw the last one. That was a terrible
way to be. Slowly I learned the biohacking techniques that enabled
me to escape my trap. I built up my strength and cardiovascular
fitness. I enhanced my metabolism and cleared away my brain fog.
I purged my stress. Above all, I learned to be vulnerable, which is
essential before you can fully relax, be kind, and be at peace. A huge
amount of my spiritual progress came from applying techniques to
unwind from the unconscious injuries and triggers that had dis-
torted my views on reality.

The psychiatrist and author Dr. Daniel Amen has developed a
test in which he shows people pictures of happy faces and angry
faces, then measures how quickly they respond.[1] To this day, my
nervous system picks out the angry faces about four times faster
than average. It's entirely unconscious. I'm programmed with threat
detection responses that are embedded so deeply in me that I cannot
hack them away. It's possible that you are built this way, too. If not,
you surely have other psychic wounds that are holding you back,
just as surely as you have personal quirks in your locomotion, your
physiology, and your metabolism. Those things help define who you
are, but they do not have to determine who you will be and what
you will do.

I often repeat the credo of self-determination: "You do you." Biohacking should set you free to figure out which "you" you want to
be—to create the best version of yourself. You have a path that got
you here. Now you can take control, sweeping away obstacles, getting the right resources, choosing your targets and pursuing them.
You have biological and spiritual reset work to do. When you peel

back enough of that stuff, when you get rid of the distractions and turn off the counterproductive alerts, you are likely to find capabilities and scars you never knew you had or that you forgot about a long time ago. I'm a very different person than I was twenty years ago. You may be shocked by how much you've evolved, and you may be shocked again by how much further you can go once you get rid of the things that are holding you back.

RECOGNIZE YOUR LENS ON REALITY

The state of your brain and body determines the version of reality you experience. If you want to get really mystical about it: reality is not real. Remember that all of your body's sensors—your cells and the 10 million billion mitochondria inside them—sense reality before you consciously do. You are always a third of a second behind, chasing them like a dog chasing a car.

The mitochondria are essentially captive bacteria living their independent lives inside you; they evolved separately from the cells they inhabit and then merged with them billions of years ago. They have their own little operating system, and they filter what they see from their own energy-making perspective. Mitochondria filter other systems in the body, including their own signals to your nervous system and your brain. Then your brain has to take an input and decide where to direct it. If it's a signal about fear or a stressful environment, it may go straight to the amygdala, bypassing your conscious prefrontal cortex entirely. You then react without thinking, following a set of rules encoded in your MeatOS.

Your lens on reality starts forming even before your prefrontal cortex is fully baked. When you are a baby, in your first month of life, your brain is dominated by delta waves, very-low-state brain waves that organize a simple view of reality, such as what a face looks like. Early childhood is almost a state of dreaming that you slowly come out of after about the age of seven. It's not until around age twenty-four that your prefrontal cortex is fully formed and programmed.

This delayed development process, which is unique to humans, means that you collect more than two decades of experience that are filtered through a preadult brain. All of your emotions, all of your youthful hurts and anxieties, shape the way you view the world.

These ideas are based on work by Jeff Hawkins, the neuroscientist and electrical engineer who created the PalmPilot, the first mobile communication device.[2] He realized that the brain constantly (and without your knowledge) makes forecasts about what will happen a microsecond in the future. It sets expectations based on those forecasts, and then it lets you notice things that don't meet the prediction. If you were to pick up your car keys tomorrow and they weighed just a little bit more than they normally do, you'd probably notice that they were heavier, because it didn't match your brain's prediction. Otherwise you wouldn't think at all about the action of moving your hand. You wouldn't be aware of picking up the car keys, and you wouldn't look at them and examine them because they met your expectations for reality.

What Jeff is describing is another aspect of the laziness principle. To maximize efficiency, your brain calls your attention only to the things it considers surprising. The rest—probably 99 percent of what is going on around you—is hidden from your conscious mind. Combine that filtering with the other forms of filtering due to childhood traumas and by your mitochondria, and it leads to some pretty weird insights. Your lens on reality is based on what your nervous system, your cells, and your body believe might be threats. In adulthood, your lens on reality is colored by your experiences and your learning, but you never lose the automated lens on reality that is based on pure programming.

SEE WHAT YOU DON'T SEE

This is where our great spiritual challenge comes from. The process of gaining awareness and kindness is actually a process of understanding that your lens on reality may not be accurate.

You may not see the same things that others see. Here's an example. Back when I was first starting my blog, I read about an urban escape and survival course, what I would call a spy school, in a book called *Emergency: This Book Will Save Your Life*.[3] It was written by Neil Strauss, who is now a friend and one of my favorite authors. He wrote about going to that spy school, which was taught by bounty hunters and military operatives. They taught how to escape from handcuffs, how to pick a lock, how to know if you're being followed, how to follow people, and so on. Then, after two days of training, the teachers would "kidnap" you with a hood over your head and handcuffs and drop you into the middle of nowhere. You would have to pick the handcuffs, escape from the kidnappers, and hide from bounty hunters while you ran a secret mission based on instructions that you receive.

I thought: Sign me up! I wanted to explore my sense of safety and vulnerability, so I took the class. One of the skills the spies taught us was how to blend in so that you can make yourself unseeable in a crowd. Now, I'm not an easy guy to hide; I'm six foot four. I literally stand out from a crowd. Yet I managed to make myself disappear.

What I did was, I put on a pair of cheap sunglasses, a red knit hat, and a fake ponytail. Then, carrying an unlit cigarette, I walked around as if I were jonesing from meth. I was shaking and shuffling as I walked down the middle of a mall in Santa Monica. Passersby carved out a ten-foot space around me because I was an untouchable, unseeable human. I walked right past three bounty hunters who were looking for me. They couldn't see me because I didn't show up in their narrow lens on reality. But there was a camera crew from A&E shooting that particular class, and those folks have a totally different lens on reality. One of the camera guys looked up, saw me, and instantly said, "Oh, look, there's Dave!" He was trained to identify anything that would make for good television, so he could see things that other people didn't, that even the bounty hunters didn't.

The bottom line is that some people can see things that other people can't. They don't just have a different attitude toward the things they see, they see entirely different things. Or they can sense entirely different things.

You may find this next story hard to believe, but it was related to me by a good friend who was in the US Army long-range reconnaissance patrol. These are soldiers who operate in the jungle for long periods of time, and their goal is to be invisible. My friend was telling about a time when he was out at night in an area where it looked as though a drug deal was going down. He realized that he might be walking into trouble. "And then, sure enough, someone had me in their sights," he said. That made no sense to me. How can you know that someone has you in their sights, if you can't even see the person? "Oh, of course, you can't see it, but you can feel it. We learn that in the military," he said. He described it as having a burning sensation in your chest and the hairs on the back of your neck standing up. When you feel that sensation, you normally dive out of the way. In his case, my friend realized that the drug dealers probably thought he was a cop, so he just stopped, lit up a big joint, and they didn't bother him anymore.

Since then I've heard similar stories from other people in the military. In those circles, it is common knowledge that you can be trained to sense a gun being aimed at you. How is it that army patrols can sense it but you can't? It's because your lens on reality has not been trained to be aware.

The process of spiritual growth requires its own form of reality training. The first thing you need to do is learn how to take your body out of the fight-or-flight mode, which causes you to overreact to preprogrammed inputs but be oblivious to things that you aren't programmed to see. All of the basic biohacks in this book will do that for you, one way or another. A course of REHIT exercise will do it. Red-light stimulus or vibration training will do it. Or you might go to a functional movement expert who will help you learn to mobilize your leg in a new way so that you can get rid of a weird movement pattern. That pattern might have been invisible to you for years, because that's how you've always moved. But an expert can take one look and see the specific muscles that you haven't been using. Sometimes an hour of therapy is all it takes to give you access to a new control system, one that was always present but was never activated.

How is it that the cameraman could see me but the bounty people couldn't? It's all about reality training. Something else interesting happened while I was hiding in plain sight in Santa Monica. All the adults saw me and thought, "Oh, geez, a homeless person on drugs. I'm outta here." But one five-year-old girl saw me, walked right up to me, smiled, and said, "Hi!" I like kids, and kids generally like me. She didn't have the reality filters telling her that I was bad and invisible, so she could see that I was just a normal, friendly person. Her mom snatched her away instantly, of course. I took my sunglasses off and said, "Don't worry. I'm not a bad guy."

That's the power of your automatic filtering system. It is part of the MeatOS, designed to keep you alive and sane. Without it, you would be flooded with too much information. But if you let it, it will keep you locked down at your lowest, laziest, least rewarding level of awareness.

Historically, changing your lens on reality and reaching a higher level of enlightenment required extreme commitment: ten years in a monastery, perhaps, or ten years fasting in a cave, or training with a shaman in the jungle while taking mind-altering substances on a daily basis. I've explored that path a little. I've had the great fortune to go to Tibet and South America and to study from different lineages to learn different approaches to spiritual growth. What I have come to appreciate is the value of traditional meditation. But the biohacker in me says that if you can use technology to accelerate your path to a more aware, more conscious state, there's great value in that, too, even without all of the learning.

The important goal is to be able to remove the bad lenses and filters on reality that aren't serving you well—that are preventing you from being kind and forgiving. With the rapid sharing of information over the internet, nearly everyone now has access to knowledge about mystical practices. With the ability to measure and monitor the alternate realities we keep in our heads, we can enter spiritual states faster and more effectively. And with the guidance of health trackers, EEG, and other feedback devices, we can travel more quickly along the long, meandering path full of suffering that's required to reach a higher level of awareness.

We have a moral imperative to use technology to improve ourselves as human beings. Military leaders do not hesitate to use technology to make people stronger, faster, and better at fighting. Shouldn't we be at least as committed to using technology to make ourselves more intelligent, thoughtful, and compassionate?

FIND YOUR HIGHEST PURPOSE

At the beginning of this book, I identified equanimity—moral serenity—as the highest state of being, the spiritual target to strive for. There are many paths that can take you there. Everything that increases your strength, resilience, energy, and clarity while reducing your stress level will make it easier for you to take that path. Those are your inputs. Acting with kindness and forgiveness spreads goodness into the world and into yourself. Those are your outputs. Equanimity just *is*. It is the state that embodies all of those other things.

In the teachings of Buddhism, there are three levels that lead up to this most elevated state.

- The first and lowest level is *empathy*, where you can imagine other people's feelings as your own. The problem with empathy is that it means you have to feel everyone else's pain and emotions. The good thing is that it will stop your feeling greed, envy, and jealousy because you are able to put yourself in the other person's shoes.
- The second level is *compassion*, where you don't have to feel the other person's pain. You can truly and deeply wish that person well, even without sharing in his or her emotions.
- The final and highest state that you can achieve is *equanimity*. You could also call it resilience. Having equanimity means that you are entirely unbothered; unruffled by whatever's happening in the world around you. This is the monk who can meditate in the middle of a storm and stay in the state of his choice, regardless of the reality

around him. This may sound like detachment, but it is more like the opposite. With equanimity, you can stay true to your principles and remain tethered to your best self even in the middle of a personal crisis—or a global pandemic.

If you follow the guidance in this book, you will have the raw materials to reach this level. Then with the right tools and the right hacks, you can add incredible levels of resilience and get a lot closer to reaching equanimity than you could any other way.

THE TRUE MEANING OF "YOU DO YOU"

We may all be striving toward the same elevated state, but we have an endless variety of styles and goals. There is no right or wrong here. When we try to create a set of rules that everyone has to follow, it always fails. It fails from a spiritual growth perspective, from an exercise and therapy perspective, and even from a nutritional perspective. It's wrong to say that a certain biohack will always work in *this* exact setting, on *this* exact schedule, for everyone. Our experiences are different, our physiologies are different, and our desires are different, too.

It's okay to say "I feel no desire to do plant medicine, even though all my friends are doing it. I'm going to try holotropic breathing instead." That's you doing you. The same thing goes for our sexual preferences and all of the other lifestyle choices that people make. If you feel as though something is wrong with you because you don't like what other people like, that is an old, counterproductive filter that you want to get rid of. You should always feel free to try something new, whether it is a supplement, a style of eating, a style of exercise, or a style of meditation. And if it doesn't work for you, you should always feel free to reject it and move on. You gave it a shot. Then you evolve and do something else until you find what works.

With modern biohacking tools, there are ways to radically change the signal that goes into your body so you can rapidly experience new states, whether it's a state of hypertrophy, putting on muscle, or a state of extreme calm and oneness with the universe. You can run experiments, pick and choose your signals, and tweak them to your own specifications, and do so more quickly than was ever possible before. Taking control of your MeatOS allows you unprecedented access to "You do you." To take advantage of that freedom, though, you have to decide on the endpoint of your path.

Making that choice requires a certain level of enlightenment. If you let your stupid programming set you on a path, you will probably do something like what I did when I was younger. I thought to myself, "I'm motivated by freedom. Money equals freedom. Therefore, I'm going to put all of my time and energy and efforts into making money." It worked—sort of. I made six million bucks when I was twenty-six. I lost it when I was twenty-eight. I told a friend of mine, in all seriousness, "I'll be happy when I have ten million dollars." I was following a goal set by my programming: work hard, obtain measurable rewards, get ahead of other people.

When you peel away all of the layers, the ultimate goal of everyone is to be happy. Money helps a little, because you need enough to eat and live comfortably, but beyond that it doesn't increase happiness. I remember visiting Cambodia in 2004, while the country was still recovering from the Khmer Rouge. Many of the people I interacted with had witnessed terrible atrocities, even seen their parents or children killed. The country was so poor that a dollar or two a day was considered a good income. Yet I saw far more happy people, sometimes walking and singing, than in my work-grind world of Silicon Valley. I realized that many of them were happier than I was. The visit opened my eyes to the vast disconnect between pursuing things we imagine will make us happy and pursuing happiness itself. People say all the time "I'll be happy when I get this thing." When they finally get it, they are happy for about ten minutes and then are profoundly disappointed to discover that the thing (whatever it is) wasn't the answer to their lack of happiness after all.

| THE HAPPINESS HACK |

The search for happiness leads straight back to the definition of bio-hacking: you change the environment around and inside yourself so that you have full control of your operating system. Happiness is a biological state. If you create the level of peace in your environment that makes you happy, it may shift over time. You can create the level of peace in yourself that makes you happy. But to do that, you have to embrace the idea that the goal is actually just to be happy, regardless of your achievements or possessions. That can be a very disorienting realization. We don't know how to quantify happiness. We don't know what's going to make us happy.

I'll share a bit of hard-earned wisdom: *you* are the only person who can make you happy. That's why spiritual growth is so important. You can learn from masters, from technology, from reading, from courses in different techniques to put yourself into different states. But at the end of the day, it all comes down to your own effort.

Happiness is a state that you can learn to induce. If you make happiness your ultimate goal, above all the specific targets you set for hacking into your MeatOS, your path will keep changing. You will find that along the way, things you think are going to get you where you want to go actually get you only halfway there. Then you adjust your goals, redirect, and keep going. The psychologist Erik Erikson described this in terms of the stages of adult development. When you're young, you focus on building your community, your tribe. Then you focus on building relationships and building your career. As you move on from there, you look at building a family, either a biological family or a group of very close people that are part of your chosen family. Then, as you get older, you turn your attention to other ways of achieving meaning, such as service to others. The things that make you happy and the things that you can do to bring happiness into the world change over time.

You can embark on the journey to happiness without upgrading your body, but it's a lot harder that way. I had a friend named

Sean Stephenson[4] who had a genetic condition called *osteogenesis imperfecta*, or brittle-bone syndrome. He was a therapist and motivational speaker who became known as "the Three Foot Giant." He was so fragile that if he sneezed wrong, he could crack a rib. He broke bones in his body more than two hundred times. Sean passed in 2019 from a wheelchair injury. Before he succumbed, one of our mutual friends asked him, "Does it upset you that this happened to you?" He said, "No, this didn't happen to me. This happened *for* me." Sean had chosen happiness despite profound physical pain and what most people would consider a horrible disability. So yes, you can be happy regardless of your physical state. But it's profoundly difficult to summon the energy Sean had unless you're an extraordinarily spiritual person.

It's just easier to be happy when your body works well. Now that you know how to control your MeatOS and bend the laziness principle to your will, you can direct some of your newfound energy into the goal of achieving happiness. It's possible to have a rock-hard body, be overflowing with energy, and still be pissed off at the world. That's why emotional and spiritual development is so important. Energy will enable you to pursue happiness, but energy alone will not take you there. You have to make the choice and commit to being your best self. Then you will become part of a great, collective upgrade.

EVALUATE, PERSONALIZE, REPEAT

Congratulations! If you've made it this far and you've invested in the process outlined in this book, you've taken the essential steps to redirect your inner laziness and make your MeatOS do the things you want it to do. You've absorbed the fundamental lessons of "smarter-not-harder." You removed friction from your life; you stocked up on the right raw materials; you picked your most important target; you figured out the right hacking signals for you; and you learned to master the art of recovery. Damn, you're good! I hope you also remembered that you don't need to be perfect at any of these things. You can't be, since you're only human. What really matters is that you are improving in ways that are meaningful to you so that you have the energy and clarity to be way, way better than normal.

But *are* you improving? Subjectively, it's easy to tell if you woke up feeling happy and motivated or if you accomplished something great today. Your friends might tell you that you're kinder than before. You might feel more connected to the people around you. You might have energy to do things you've been meaning to do for years. These are all signals that yes, you are moving in the right direction.

Still, we all have bad days that make it difficult to be sure we are on the right track. More important, we can easily fool ourselves into thinking that something is working even when it isn't. I spent years imagining that painful gym workouts were making me strong and that a vegan diet was making me healthy. We biohackers don't want to rely on self-impressions alone. We certainly don't want to rely on authorities telling us, "Trust me, I'm sure it works."

When I say that you shouldn't trust anyone on faith alone, that includes me. Don't accept all of my insights into hacking your MeatOS just because I say they work. Trust is an essential starting point, but you want to measure and verify. You also want to make sure that you're applying the hacks in the best way for *you*. Everyone's biology is a little different, for starters. Then there are the specifics of your environment and your lifestyle. Also, life is not static. Each day your situation changes, and your goals shift constantly as well. It's crucial to evaluate your hacks; personalize them so you get as much out of them as possible; and repeat the process on a regular basis.

Fundamentally, you have to know yourself if you want to become a better version of yourself. This is the final step in the six-step process of biohacking I spelled out in chapter 1.

Evaluate Your Setup

Look back and judge how well you're hacking your laziness.

Step 1: Remove friction.
Step 2: Load up on raw materials.

All of your work to upgrade your MeatOS will go nowhere if you didn't address these two important steps that set you up for success in hacking your biology.

How well are you doing at these? You can take objective measures of your inputs—the specifics of what you are eating and how you are living—and then take objective measurements of your outputs, or how your body is responding to the changes you have made. The best way to do that is by running medical tests.

- You can test your mineral status by getting a red blood cell mineral profile, which will show the status of your macrominerals (and some of your microminerals). If you see a doctor, you can ask him or her to order lab testing for you, or you may be able to order a test online, depending on where you live. A doctor will draw your blood and evaluate your mineral status. Your body replaces all of your red blood cells every four months, so your blood test will give only a brief history of your supply of essential resources, but it can give you a general idea of how you are doing mineral-wise.

- You can also ask your doctor for a trace mineral panel that will measure chromium, copper, iodine, iron, manganese, molybdenum, selenium, and zinc. Another easy way to test your mineral status at home is with a hair mineral testing kit. Although this method is controversial, there are more than forty years of research behind hair mineral analysis. Studies show that dietary levels of microelements correspond to the concentrations of those elements in hair, and a hair mineral analysis will give you an estimate of your mineral status over a period of time.[1]

- You can ask your doctor for a vitamin D blood test, or you can get an at-home testing kit. Remember, you want to aim for levels between 70 and 90 ng/mL. Your levels of vitamins A and E can also be assessed with blood tests.

- You can test your vitamin K1 status with a blood test that measures your blood clotting, called a prothrombin time test, but there are no accessible tests to measure your K2. You can, however, test your arterial calcification level[2] with a coronary artery calcium score, and if you remember from chapter 4, vitamin K2 helps keep calcium in your bones and out of your arteries.

The simplest way to know whether or not you're improving on your raw material status is to do a physical self-evaluation. Do you have more energy throughout the day? How do your hair, skin, and nails look? If you're seeing improvements in these areas, this is a good indicator that you have made positive changes. If not, you want to do some fine-tuning.

Evaluate Your Choices

Are you happy with your choice of a target area to go after? Go back to page 126 and run your numbers. See if the result feels right to you. Will pursuing this target make you an overall better person and help you reach your life goals? What changes will you be making? Can you see yourself doing them? Is there a way to measure your progress?

Evaluate Your Outcomes

This is where the rubber hits the road—or rather, where the signals meet the MeatOS and you find out if you are really changing and improving the way you intended. This is your chance to get past the "Did I just imagine it?" questions and see exactly what slope-of-the-curve signals can do for you.

When evaluating your hacks, be sure you take a baseline, or "before," measure and as many measurements as possible throughout the experiment. If you don't have a baseline to compare your results with, you won't truly know if things are improving.

There are three ways you can measure and monitor the results of your hacks.

- You can perform subjective daily monitoring of things such as sleep quality and stress levels. This is the simplest way to track your results because it doesn't require any fancy tech. All you need is a place to store your thoughts. A simple daily note of how you are feeling on a scale of 1 to 10 as soon as you wake up can be an incredibly powerful way to track how you're doing over a period of time.
- You can use real-time tracking devices such as a fitness tracker or continuous glucose monitor. These devices give you objective and quantitative measures of how you're doing. They are sometimes inaccurate, but they are a great method of establishing a baseline and tracking trends over time.

- You can perform occasional or yearly deep lab tests or imaging. These are more expensive and intrusive, but they give you a deep, objective picture of the state of your biology. Extensive bloodwork, a coronary artery calcification scan, or a full-body MRI are examples of such tests.

MEASURE YOURSELF AGAINST YOUR SPECIFIC TARGETS

The more focused you are with your goals, the more precise you can be in determining whether you're on the right path.

Cardiovascular Fitness

You can assess your cardiovascular fitness by measuring your resting heart rate and your VO_2 max. If you have a smart watch or other heart rate–tracking device, you can use it to track your resting heart rate. Generally, a lower resting heart rate correlates with better cardiovascular fitness. Measuring VO_2 max is more challenging. To get a high-accuracy assessment, you need to be evaluated in a laboratory setting, which most people can't do on a regular basis. You can get a rough estimate, though, by using an online calculator that asks you to input parameters such as your age, height, and maximum heart rate. You can easily find one by searching "VO_2 max calculator" on a trusted search engine. A higher VO_2 max reading is associated with greater cardiovascular fitness.

Strength and Muscle

You can measure your improving strength by testing your one-rep max at the gym—if you are still bothering to go to a gym, that is. Be sure to use a machine or have a spotter available if you're using free weights, so that you don't injure yourself. A dual-energy X-ray absorptiometry (DEXA) scan (more expensive and less accessible) or bioelectrical impedance analysis (more accessible) will determine your body fat percentage and can help you track muscle gains and fat loss over time. You can find a DEXA scan provider near you by conducting an internet search. Many fitness centers, gyms, and chiropractors have bioelectrical impedance assay equipment that staff members can guide you through using. Even if you don't have access to these, you can take your arm, leg, and waist measurements with a measuring tape to get an estimate of your muscle gain and fat loss progress.

Energy

For energy, the measure is simply: How do you feel? Could you get up and dance right now? Could you do a full-on sprint? If you have more energy, you will feel it.

Brain

You can measure your brain performance using an EEG test or a working memory test. If you have the means, you can go for a single-photon emission computerized tomography (SPECT) scan like the one I got at Daniel Amen's clinic. A SPECT scan isn't cheap, but it will generate a 3D snapshot of your brain that can tell you which areas are working well and which ones aren't. For the rest of us, a memory test or reaction time test are useful in assessing cognitive function progress.

Stress

For stress and anxiety, you can measure your heart rate variability, or HRV. When I started out as a biohacker, you generally had to go to a doctor's office to get an EKG in order to measure your variability. Today, you can easily buy home HRV monitors, and even some watchlike fitness trackers estimate this number for you. In general, a higher level of HRV means that you are more flexible and adaptable, and therefore less stressed.

Sex

For sex, ask your partner—or yourself. How is your sex drive? If you're a guy, a good way to figure out if your hormones are in the right place (or not) is to check whether or not you wake up with a morning—ahem—kickstand. You can also run a blood test to monitor your hormones. A full hormone panel from a doctor can help you figure out whether you're low or high on certain hormones that impact your sexual function such as estrogen, progesterone, testosterone, thyroid hormones, and cortisol.

Sleep

Monitoring how *long* you sleep doesn't tell you how *well* you sleep. If you have a sleep tracker, you can dig into the more revealing data: how long it took you to fall asleep, as well as how much time you spent in deep and REM sleep. About 1.5 hours each of deep sleep and REM sleep during the night is ideal. Keep a journal of how rested you feel when you wake up and what your fatigue levels are like throughout the day; these self-evaluations are also good indicators of your sleep quality.

Longevity

Testing longevity can be tricky, unless you've found a way to peer into the future. A simple way to determine how well you're aging is to look in the mirror—but that can be too simple and too subjective. Better, take a DNA methylation test, which measures your epigenetic age based on the pattern of methyl molecule fragments that are attached to your DNA. These fragments switch bits of your genome on or off, and they change as you get older. DNA methylation is currently one of the most useful and accurate ways to measure our biological aging processes.

As you use these various objective measures to evaluate yourself, keep in mind that the raw numbers are not the whole story. Yes, they are helpful for validating that your hacks are working and for understanding how you personally are responding to them. It's important to remember, though, that your subjective sense really matters. If you are targeting your brain, for instance, EEG readings are helpful, but what you really care about is: Do you struggle to find the right words? Do you open the refrigerator or go to the store and not remember what you were looking for? Those are common problems, and they are the biggest signs that you have a brain issue, regardless of your EEG results. If they go away, you will know that your hacks are working.

| PERSONALIZE YOUR HACKS |

Along my biohacking journey, I have seen over and over that what works for one person often does not work for another. If you aren't quite getting the results you were aiming for, that's okay. Part of biohacking is making mistakes and learning what to do if something doesn't work or give us the outcome we had expected.

Before throwing something away and deeming it useless, though, it is important that you give your hacks enough time to work. Upgrades and biohacks are often shortcuts for those of us who want more results in less time, but sometimes we need to maintain a bit of patience. Remember, you didn't gain those extra fifty pounds in a day, so you sure as heck aren't going to lose them in a day, either.

It is also important that you be consistent. You can't do one workout and expect it to solve your problems. You have to show up regularly to get results. If you are having trouble with consistency, find someone who can hold you accountable.

That all being said, you don't want to waste your whole life doing things that aren't working, hoping that one day they'll magically start working for you. To avoid dead ends, I go at biohacking with the approach "Throw everything at the problem that could possibly work. If it really does work, great!" That's the smarter way. Once I get a positive result, I can sort out the details and remove items one by one to decide which things make me better and which don't really make a difference. Otherwise, the process would be agonizingly slow.

| IMPROVE CONTINUOUSLY |

Don't forget to upgrade your upgrades. Our bodies are constantly changing, which means that the biohacks that work best for you today might not be the hacks that will work the best for you two months from now. Your goals and circumstances will change. That's not just inevitable, it's desirable. In an active, well-lived life, you will be constantly evolving and exploring. Science and technology are advancing rapidly, too, bringing new possibilities to the biohacking table. There is always room for improvement.

It is amazing that the human body—the thing we should know better than anything else in the world—is still full of mystery and untapped potential. Embrace this marvel and take the "repeat" part of "evaluate, personalize, repeat" not as a slog but as an extraordinary opportunity to discover unexplored parts of yourself.

AN OFFERING TO THE WORLD

Every day, everything you do is an offering to the world. You are putting a little bit of yourself out there, making things slightly better (or slightly worse) as a result. This book is one part of my offering to the world. It contains the best techniques I know for taking control of the mindless operating system inside of you and making yourself better than normal. With you in charge of your MeatOS, you can redirect your biological laziness so that it will make you strong. Then you will face the most profound challenge in life: What should you do with your energy? How can you become a kind, resilient, happy, more fully realized version of yourself?

Imagine how much easier it would be if every person came with a control, like a navigation screen in your car, that would let you punch in a destination and then map out the best way to get there. You could set your speed, track your progress, figure out if you have enough energy to get there, know when you will arrive. If only. On the spiritual and emotional level, just as on the biological level, our operating system is invisible to us. We don't have the instrumentation that lets us say "Ah, yes, that spot right there. That's who I want to be and what I want to be like."

The whole point of biohacking is to let us start to create those kinds of instruments, at least rudimentary versions of them. Biohacking draws together the best ideas from history, including thousands of years of spiritual practice and even older traditions of nutritional and exercise practices that are as old as our species. On top of that we add a century and a half of psychology and biochemistry. Then come layers of genetics, epigenetics, proteomics, nootropics, brain mapping, artificial intelligence . . . we're doing what people have been trying to do for hundreds of generations, but we can do it a lot faster and more strategically because we now have an unprecedented ability to measure and crunch data about ourselves.

Oh, and we can share information. If the guys meditating in the Himalayas long ago had a great insight, how were they going to tell the guys meditating in the Andes? Make a connection in the dream world, maybe. Today, biohacking information is online for everyone to access.

What you do with that information is entirely up to you. Maybe your immediate targets are to be lean and muscular and look a few years younger. There is no shame in that. If you feel better about yourself, you will be happier and kinder. Biohacking is not an *either/or* process, it is an *and* process. All the ways that you upgrade yourself will support one another. The more you are in charge of your MeatOS, regardless of your target, the better the offerings you can bring into the world.

Right now, at Upgrade Labs, our team is rolling out technology to take people on guided neurofeedback journeys. We want to help make spiritual resets easier and more accessible, just as we are making exercise and physical resets more accessible. At every level, the biohacking philosophy is the same: life is too short to waste it on outdated ideas—working hard, rather than working smart—that don't lead to genuine self-improvement and it is too precious to settle for being a baseline "normal" version of yourself.

When you push hard against the slope of the curve, you will be amazed by how far you can go in redirecting the laziness programmed into your biology. You will experience new levels of where your mind and body can go and discover just how lofty the calm, "normal" version of you can be.

ACKNOWLEDGMENTS

Welcome to the gratitude section. Because you've completed the book, you already know how important gratitude is to your own happiness and performance. That doesn't mean it's easy to express, because there are so many people I am grateful for. I'm grateful for the people who have listened to *The Human Upgrade*™ (formerly *Bulletproof Radio*). I'm grateful for the people who stop me at conferences, at events, in restaurants, and at the grocery store to tell me the changes they've made from listening to the show or reading my books. It's those kinds of interactions that inspire me to put in the work to write a book like this and to host multiple podcast episodes a week.

Many thanks to my writing team; Corey S. Powell, my writing partner; Julie Will, my editor; and Celeste Fine, my agent. Words do not express how grateful I am for your time and energy. A very special thanks to Nicole Peterson for bringing her knowledge and expertise in the world of biohacking. Super-thanks also to Christine Tenove, my assistant, who manages my packed calendar to ensure I hit my deadlines so I can be a father, CEO, author, and podcaster, and still have time to biohack and practice self-care.

Speaking of time, a big thank you to my family, Lana, Anna, and Alan; I am grateful for their support, patience, and love. They know that the insights they share are guiding multiple generations. And special thanks to my teams at TrueDark/TrueLight, 40 Years of Zen, Homebiotic, Danger Coffee, The Upgrade Collective, and Upgrade Labs.

I've been biohacking for well over a decade, always seeking ways to work smarter, not harder. Along the way, so many new hacks have come out, and I am grateful to have had the opportunity to talk to many of the world's top leaders who have guided me. Special thanks to Dr. Daniel Amen, who helped me get my brain in order years ago, and to my dear friend Mike Koenigs for his continued guidance. Thanks also to all of my spiritual teachers and masters who have supported me and educated me on the path to creating this book. You know who you are.

Finally, I am grateful to a few friends and groups who have shared extra business support and wisdom: Joe Polish's Genius Network, J.J. Virgin's Mindshare Group, Michael Fishman's Consumer Health Summit, GoBundance, and YourBestLife.

Assuming you are reading this, I would like to thank you, too, for investing your time and attention in this book. I sincerely hope it was more than worth the effort you put into it.

Welcome to the new you!

NOTES

INTRODUCTION: BETTER THAN NORMAL

1. Elizabeth Pegg Frates, "Did We Really Gain Weight During the Pandemic?," Harvard Health Publishing, October 5, 2021, https://www.health.harvard.edu/blog/did-we-really-gain-weight-during-the-pandemic-202110052606.

1: TAP THE POWER OF LAZINESS

1. Robert J. Kosinski, "A Literature Review on Reaction Time," Clemson University course materials, September 2013, https://www.fon.hum.uva.nl/rob/Courses/InformationInSpeech/CDROM/Literature/LOTwinterschool2006/biae.clemson.edu/bpc/bp/Lab/110/reaction.htm.
2. "Blue Box, Designed and Built by Steve Wozniak and Marketed by Steve Jobs, Circa 1972," The Henry Ford, https://www.thehenryford.org/collections-and-research/digital-collections/artifact/452666/.
3. Joseph Pizzorno, "Mitochondria: Fundamental to Life and Health," *Integrative Medicine* 13, no. 2 (2014): 8–15, https://www.ncbi.nlm.nih.gov/pmc/articles/PMC4684129/.
4. "More than 73% of American Adults Are Overweight or Obese," Diabetes.co.uk, December 29, 2020, https://www.diabetes.co.uk/news/2020/dec/more-than-73-of-american-adults-overweight-or-obese.html.

2: REMOVE YOUR FRICTION

1. "Are Anti-nutrients Harmful?," Harvard T. H. Chan School of Public Health, https://www.hsph.harvard.edu/nutritionsource/anti-nutrients/.
2. Ellen C. G. Grant, "Rapid Response to: 'Operative Delivery and Postnatal Depression: A Cohort Study,'" *BMJ* 2005 (330): 879, https://www.bmj.com/rapid-response/2011/10/30/zinc-and-copper-deficiencies-can-cause-postpartum-depression.
3. "Lectins," Harvard T. H. Chan School of Public Health, https://www.hsph.harvard.edu/nutritionsource/anti-nutrients/lectins/.
4. WebMD Editorial Contributors, "What Is Oxalate (Oxalic Acid)?," Web MD, April 8, 2021, https://www.webmd.com/diet/what-is-oxalate-oxalic-acid.

5. David L. J. Freed, "Do Dietary Lectins Cause Disease?," *BMJ* 308, no. 7190 (1999): 1023–24, https://www.ncbi.nlm.nih.gov/pmc/articles/PMC1115436/.

6. "The Bulletproof Guide to Omega 3 vs. Omega 6 Fats," Bulletproof, April 28, 2022, https://www.bulletproof.com/supplements/aminos-enzymes/omega-3 -vs-omega-6-fat-supplements/.

7. James J. DiNicolantonio and James H. O'Keefe, "Omega-6 Vegetable Oils as a Driver of Coronary Heart Disease: The Oxidized Linoleic Acid Hypothesis," *Open Heart* 5, no. 2 (2018): e000898, https://openheart.bmj.com/content /5/2/e000898.

8. Sara Huerta-Yépez, Ana B. Tirado-Rodriguez, and Oliver Hankinson, "Role of Diets Rich in Omega-3 and Omega-6 in the Development of Cancer," *Boletín Médico del Hospital Infantil de México* 73, no. 6 (2016): 446–56, https://www.sciencedirect.com/science/article/pii/S1665114616301423.

9. E. M. Sullivan et al., "Murine Diet-Induced Obesity Remodels Cardiac and Liver Mitochondrial Phospholipid Acyl Chains with Differential Effects on Respiratory Enzyme Activity," *Journal of Nutritional Biochemistry* 45 (2017): 94–103, https://pubmed.ncbi.nlm.nih.gov/28437736/.

10. Rekhadevi Perumalla Venkata and Rajagopal Subramanyam, "Evaluation of the Deleterious Health Effects of Consumption of Repeatedly Heated Vegetable Oil," *Toxicology Reports* 3 (2016): 636–43, https://www.ncbi.nlm .nih.gov/pmc/articles/PMC5616019/; Maria D. Guillén and Patricia S. Uriarte, "Aldehydes Contained in Edible Oils of a Very Different Nature After Prolonged Heating at Frying Temperature: Presence of Toxic Oxygenated α,β Unsaturated Aldehydes," *Food Chemistry* 131, no. 3 (2012): 915–26, https://www.sciencedirect.com/science/article/abs/pii/S03088146110135 62?via%3Dihub; Marni Stott-Miller, Marian L. Neuhouser, and Janet L. Stanford, "Consumption of Deep-Fried Foods and Risk of Prostate Cancer," *Prostate* 73, no. 9 (2013): 960–69, https://onlinelibrary.wiley.com/doi/10.1002 /pros.22643.

11. Christopher E. Ramsden et al., "The Sydney Diet Heart Study: A Randomised Controlled Trial of Linoleic Acid for Secondary Prevention of Coronary Heart Disease and Death," *FASEB Journal* 27, no. S1 (2013): 127.4, https://faseb .onlinelibrary.wiley.com/doi/abs/10.1096/fasebj.27.1_supplement.127.4.

12. James J. DiNicolantonio, "The Importance of Maintaining a Low Omega-6/ Omega-3 Ratio for Reducing the Risk of Inflammatory Cytokine Storms," *Missouri Medicine* 117, no. 6 (2020): 539–42, https://www.ncbi.nlm.nih.gov /pmc/articles/PMC7721408/.

13. U.S. Food and Drug Administration, "Trans Fat," May 18, 2018, https:// www.fda.gov/food/food-additives-petitions/trns-fat.

14. Jianzhong Hu et al., "Low-Dose Exposure of Glyphosate-Based Herbicides Disrupt the Urine Metabolome and Its Interaction with Gut Microbiota," *Scientific Reports* 11 (2021): art. 3265, https://www.nature.com/articles /s41598-021-82552-2; Robin Mesnage et al., "Use of Shotgun Metagenomics and Metabolomics to Evaluate the Impact of Glyphosate or Roundup MON 52276 on the Gut Microbiota and Serum Metabolome of Sprague-Dawley Rats," *Environmental Health Perspectives* 129, no. 1 (2021), https://ehp.niehs .nih.gov/doi/10.1289/EHP6990.

15. Luoping Zhang et al., "Exposure to Glyphosate-Based Herbicides and Risk

for Non-Hodgkin Lymphoma: A Meta-analysis and Supporting Evidence," *Mutation Research/Reviews in Mutation Research* 781 (2019): 186–206, https://www.sciencedirect.com/science/article/abs/pii/S1383574218300887.

16. Olha M. Stribyska et al., "The Effects of Low-Toxic Herbicide Roundup and Glyphosate on Mitochondria," *EXCLI Journal* 21 (2022): 183–96, https://www.ncbi.nlm.nih.gov/pmc/articles/PMC8859649/#R86.

17. O. O. Olorunsogo, "Modification of the Transport of Protons and Ca2+ Ions Across Mitochondrial Coupling Membrane by N-(Phosphonomethyl)glycine," *Toxicology* 61, no. 2 (1990): 205–09, https://pubmed.ncbi.nlm.nih.gov /2157305/.

18. Francisco Peixoto, "Comparative Effects of the Roundup and Glyphosate on Mitochondrial Oxidative Phosphorylation," *Chemosphere* 61, no. 8 (2005): 1115–22, https://pubmed.ncbi.nlm.nih.gov/16263381/.

19. Stribyska et al., "The Effects of Low-Toxic Herbicide."

20. Wesley R. Harris et al., "Computer Simulation of the Interactions of Glyphosate with Metal Ions in Phloem," *Journal of Agricultural and Food Chemistry* 60, no. 24 (2012): 6077–87, https://www.ncbi.nlm.nih.gov/pubmed/2265 1133.

21. Alexis Temkin and Olga Naidenko, "Glyphosate Contamination in Food Goes Far Beyond Oat Products," Environmental Working Group, February 28, 2019, https://www.ewg.org/news-insights/news/glyphosate-contamination-food -goes-far-beyond-oat-products.

22. "Potential Health Risks Associated with Stressed Foodstuffs Such as Foie Gras," ScienceDaily, February 19, 2009, https://www.sciencedaily.com/re leases/2009/02/090210092736.htm.

23. Insaf Berrazaga et al., "The Role of the Anabolic Properties of Plant- versus Animal-Based Protein Sources in Supporting Muscle Mass Maintenance: A Critical Review," *Nutrients* 11, no. 8 (2019): 1825, https://www.ncbi.nlm.nih .gov/pmc/articles/PMC6723444/.

24. Shruti Jain et al., "Tracing the Role of Plant Proteins in the Response to Metal Toxicity: A Comprehensive Review," *Plant Signaling & Behavior* 13, no. 9 (2018): e1507401. https://www.ncbi.nlm.nih.gov/pmc/articles/PMC620 4846/.

25. Allison J. Hodgkinson, Natalie A. McDonald, and Brad Hine, "Effect of Raw Milk on Allergic Responses in a Murine Model of Gastrointestinal Allergy," *British Journal of Nutrition* 112, no. 3 (2014): 390–97, https://pubmed.ncbi .nlm.nih.gov/24870507/; S. Ho et al., "Comparative Effects of A1 Versus A2 Beta-Casein on Gastrointestinal Measures: A Blinded Randomised Cross-over Pilot Study," *European Journal of Clinical Nutrition* 68 (2014): 994–1000, https://www.nature.com/articles/ejcn2014127; Daniela Kullenberg de Gaudry et al., "Milk A1β-Casein and Health-Related Outcomes in Humans: A Systematic Review," *Nutrition Reviews* 77, no. 5 (2019): 278–306, https:// academic.oup.com/nutritionreviews/article/77/5/278/5307073.

26. Ho et al., "Comparative Effects of A1 Versus A2 Beta-Casein."

27. Hodgkinson, McDonald, and Hine, "Effect of Raw Milk"; Ho et al., "Comparative Effects of A1 Versus A2 Beta-Casein."

28. Fang Qian et al., "Experimental and Modelling Study of the Denaturation of Milk Protein by Heat Treatment," *Korean Journal for Food Science of*

Animal Resources 37, no. 1 (2017): 44–51, https://www.ncbi.nlm.nih.gov/pmc
/articles/PMC5355583; Yangdong Zhang et al., "Proteomics Analysis Reveals
Altered Nutrients in the Whey Proteins of Dairy Cow Milk with Different
Thermal Treatments," *Molecules* 26, no. 15 (2021): 4628, https://www.ncbi
.nlm.nih.gov/pmc/articles/PMC8347753/.

29. Ton Baars, "Milk Consumption, Raw and General, in the Discussion on
    Health or Hazard," *Journal of Nutritional Ecology and Food Research* 1,
    no. 2 (2013): 91–107, https://www.researchgate.net/publication/255685679
    _Milk_Consumption_Raw_and_General_in_the_Discussion_on_Health
    _or_Hazard.

30. Hodgkinson, McDonald, and Hine, "Effect of Raw Milk."

31. Aiqian Ye et al., "Effect of Homogenization and Heat Treatment on the
    Behavior of Protein and Fat Globules During Gastric Digestion of Milk,"
    Journal of Dairy Science 100, no. 1 (2017): 36–47, https://pubmed.ncbi.nlm
    .nih.gov/27837978/; Michael H. Tunick et al., "Effect of Heat and Homog-
    enization on in Vitro Digestion of Milk," *Journal of Dairy Science* 99, no. 6
    (2016): 4124–39, https://www.journalofdairyscience.org/article/S0022-0302
    (16)30140-0/pdf; Bolin Mou et al., "Phospholipidomics of Bovine Milk Sub-
    jected to Homogenization, Thermal Treatment and Cold Storage," *Food
    Chemistry* 381 (2022): 132288, https://pubmed.ncbi.nlm.nih.gov/35124494/.

32. Sameh Obeid et al., "The Surface Properties of Milk Fat Globules Govern
    Their Interactions with the Caseins: Role of Homogenization and pH Probed
    by AFM Force Spectroscopy," *Colloids and Surfaces B: Biointerfaces* 182
    (2019): 110363https://pubmed.ncbi.nlm.nih.gov/31344611/.

33. Ye et al., "Effect of Homogenization and Heat Treatment."

34. Agnieszka Rogowska et al., "Zearalenone and Its Metabolites: Effect on Hu-
    man Health, Metabolism and Neutralisation Methods," *Toxicon* 162, no. 2
    (2019): 46–56, https://pubmed.ncbi.nlm.nih.gov/30851274/.

35. Herbert Hof, "Mycotoxins in Milk for Human Nutrition: Cow, Sheep and
    Human Breast Milk," *GMS Infectious Diseases* 4 (2016), https://www.ncbi
    .nlm.nih.gov/pmc/articles/PMC6301711/; Rogowska et al., "Zearalenone and
    Its Metabolites."

36. "Mycotoxins," World Health Organization, May 9, 2018, https://www.who
    .int/news-room/fact-sheets/detail/mycotoxins.

37. Hayley K. McIlwraith et al., "Evidence of Microplastic Translocation in
    Wild-Caught Fish and Implications for Microplastic Accumulation Dynam-
    ics in Food Webs," *Environmental Science & Technology* 55, no. 18 (2021):
    12372–82, https://pubs.acs.org/doi/full/10.1021/acs.est.1c02922; Madeleine
    Smith et al., "Microplastics in Seafood and the Implications for Human
    Health," *Current Environmental Health Reports* 5, no. 3 (2018): 375–86,
    https://www.ncbi.nlm.nih.gov/pmc/articles/PMC6132564/; Md. Simul Bhu-
    yan, "Effects of Microplastics on Fish and in Human Health," *Frontiers in
    Environmental Science*, March 16, 2022, https://www.frontiersin.org/articles
    /10.3389/fenvs.2022.827289/full.

38. Smith et al., "Microplastics in Seafood."

39. Fatih Gultekin et al., "Food Additives and Microbiota," *Northern Clinics
    of Istanbul* 7, no. 2 (2020): 192–200, https://www.ncbi.nlm.nih.gov/pmc
    /articles/PMC7117642/; Zhengxiang He et al., "Food Colorants Metabolized

by Commensal Bacteria Promote Colitis in Mice with Dysregulated Expression of Interleukin-23," *Cell Metabolism* 33, no. 7 (2021): 1358–71, https://www.ncbi.nlm.nih.gov/pmc/articles/PMC8266754/.

40. Y. Zhou and N. C. Danbolt, "Glutamate as a Neurotransmitter in the Healthy Brain," Journal of Neural Transmission 121, no. 8 (2014): 799–817, https://www.ncbi.nlm.nih.gov/pmc/articles/PMC4133642/.

3: LOAD UP ON RAW MINERALS

1. Albina Nowak et al., "Effect of Vitamin D3 on Self-Perceived Fatigue," *Medicine* 95, no. 52 (2016): e5353, https://www.ncbi.nlm.nih.gov/pmc/articles/PMC5207540/; Akash Sinha et al., "Improving the Vitamin D Status of Vitamin D Deficient Adults Is Associated with Improved Mitochondrial Oxidative Function in Skeletal Muscle," *Journal of Clinical Endocrinology & Metabolism* 98, no. 3 (2013): e509–13, https://pubmed.ncbi.nlm.nih.gov/23393184/.

2. Anne-Laure Tardy et al., "Vitamins and Minerals for Energy, Fatigue and Cognition: A Narrative Review of the Biochemical and Clinical Evidence," *Nutrients* 12, no. 1 (2020): 228, https://www.ncbi.nlm.nih.gov/pmc/articles/PMC7019700/.

3. Elad Tako, "Dietary Trace Minerals," *Nutrients* 11, no. 11 (2019): 2823, https://www.ncbi.nlm.nih.gov/pmc/articles/PMC6893782/.

4. Susana Puntarulo, "Iron, Oxidative Stress and Human Health," *Molecular Aspects of Medicine* 26, 4–5 (2005): 299–312, https://pubmed.ncbi.nlm.nih.gov/16102805/.

5. Kazumasa Yamagishi et al., "Dietary Fiber Intake and Risk of Incident Disabling Dementia: The Circulatory Risk in Communities Study," *Nutritional Neuroscience*, February 6, 2022, https://www.tandfonline.com/doi/full/10.1080/1028415X.2022.2027592.

6. Astrid Kolderup Hervik and Birger Svihus, "The Role of Fiber in Energy Balance," *Journal of Nutrition and Metabolism* 2019 (2019): art. 4983657, https://www.ncbi.nlm.nih.gov/pmc/articles/PMC6360548/.

7. Richard B. Kreider et al., "International Society of Sports Nutrition Position Stand: Safety and Efficacy of Creatine Supplementation in Exercise, Sport, and Medicine," *Journal of the International Society of Sports Nutrition* 14 (2017): 18, https://www.ncbi.nlm.nih.gov/pmc/articles/PMC5469049/.

8. David Benton and Rachel Donohoe, "The Influence of Creatine Supplementation on the Cognitive Functioning of Vegetarians and Omnivores," *British Journal of Nutrition* 105, no. 7 (2011): 1100–05, https://pubmed.ncbi.nlm.nih.gov/21118604/.

9. Jinmo Khil and Daniel D. Gallaher, "Beef Tallow Increases Apoptosis and Decreases Aberrant Crypt Foci Formation Relative to Soybean Oil in Rat Colon," *Nutrition and Cancer* 50, no. 1 (2004): 55–62, https://pubmed.ncbi.nlm.nih.gov/15572298/.

10. Charles M. Benbrook et al., "Enhancing the Fatty Acid Profile of Milk Through Forage-Based Rations with Nutrition Modeling of Diet Outcomes," *Food Science & Nutrition* 6, no. 3 (2018): 681–700, https://onlinelibrary.wiley.com/doi/10.1002/fsn3.610.

11. Leah D. Whigham, Abigail C. Watras, and Dale A. Schoeller, "Efficacy of

Conjugated Linoleic Acid for Reducing Fat Mass: A Meta-analysis in Humans," *American Journal of Clinical Nutrition* 85, no. 5 (2007): 1203–11, https://academic.oup.com/ajcn/article/85/5/1203/4632999.

12. Cision, "Statement—Consider Using Alternatives to Palm Supplements, Says Dairy Farmers of Canada," Newswire, February 25, 2021, https://www.newswire.ca/news-releases/statement-consider-using-alternatives-to-palm-supplements-says-dairy-farmers-of-canada-873991654.html.

13. Brad Heins, "Grass-Fed Cows Produce Healthier Milk," University of Minnesota Extension, 2021, https://extension.umn.edu/pasture-based-dairy/grass-fed-cows-produce-healthier-milk.

14. Robin Mesnage et al., "An Integrated Multi-omics Analysis of the NK603 Roundup-Tolerant GM Maize Reveals Metabolism Disturbances Caused by the Transformation Process," *Scientific Reports* 6 (2016): art. 37855, https://www.nature.com/articles/srep37855.

15. U.S. Department of Agriculture, "Artichokes (Globe or French), Raw," FoodData Central, April 1, 2019, https://fdc.nal.usda.gov/fdc-app.html#/food-details/169205/nutrients.

16. L. A. Moreno et al., "Psyllium Fibre and the Metabolic Control of Obese Children and Adolescents," *Journal of Physiology and Biochemistry* 59, no. 3 (2003): 235–42, https://pubmed.ncbi.nlm.nih.gov/15000455/.

17. University of Massachusetts Amherst, "Brassicas, Alternaria Leaf Spot," Center for Agriculture, Food, and the Environment, UMass Extension Vegetable Program, January 2013, https://ag.umass.edu/vegetable/fact-sheets/brassicas-alternaria-leaf-spot.

18. Jan Alexander, "Selenium," in *Handbook on the Toxicology of Metals*, 4th ed. vol. 2, edited by Gunnar F. Nordberg, Bruce A. Fowler, and Monica Nordberg (Cambridge, MA: Academic Press, 2015), 1175–1208, https://www.sciencedirect.com/science/article/pii/B9780444594532000524.

4: SUPPLEMENT YOUR MeatOS

1. Stephanie Seneff, interviewed by Dave Asprey, "Transcript: Glyphosate Toxicity, Lower Cholesterol Naturally & Get Off Statins," *Bulletproof* podcast no. 238, https://daveasprey.com/transcript-dr-stephanie-seneff-glyphosate-toxicity-lower-cholesterol-naturally-get-off-statins-238/.

2. Kristie L. Ebi and Irakli Loladze, "Elevated Atmospheric CO_2 Concentrations and Climate Change Will Affect Our Food's Quality and Quantity," *Lancet Planetary Health* 3, no. 7 (2019): e283–84, https://www.thelancet.com/journals/lanplh/article/PIIS2542-5196(19)30108-1/fulltext.

3. "27 Years—No Deaths from Vitamins, 3 Million from Prescription Drugs," Natural Society, October 3, 2011, last updated July 29, 2021, https://naturalsociety.com/27-years-no-deaths-from-vitamins-3-million-prescription-drug-deaths/.

4. Barbara Prietl et al., "Vitamin D and Immune Function," *Nutrients* 5, no. 7 (2013): 2502–21, https://www.ncbi.nlm.nih.gov/pubmed/23857223.

5. Dov Tiosano et al., "The Role of Vitamin D Receptor in Innate and Adaptive Immunity: A Study in Hereditary Vitamin D–Resistant Rickets Patients," *Journal of Clinical Endocrinology & Metabolism* 98, no. 4 (2013): 1685–93, https://www.ncbi.nlm.nih.gov/pubmed/23482605/; Cedric F. Garland et al.,

"The Role of Vitamin D in Cancer Prevention," *American Journal of Public Health* 96, no. 2 (2006): 252–61, https://www.ncbi.nlm.nih.gov/pmc/articles /PMC1470481.

6. S. Pilz et al., "Effect of Vitamin D Supplementation on Testosterone Levels in Men," *Hormone and Metabolic Research* 43, no. 3 (2011): 223–25, https:// www.ncbi.nlm.nih.gov/pubmed/21154195; Julia A. Knight et al., "Vitamin D Association with Estradiol and Progesterone in Young Women," *Cancer Causes & Control* 21, no. 3 (2010): 479–83, https://www.ncbi.nlm.nih.gov /pubmed/19916051.

7. Christopher Masterjohn, "Vitamin D Toxicity Redefined: Vitamin K and the Molecular Mechanism," *Medical Hypotheses* 68, no. 5 (2007): 1026–34, https://pubmed.ncbi.nlm.nih.gov/17145139/.

8. "About Vitamin D," Vitamin D Council, 2018, https://www.vitamindcouncil .org/about-vitamin-d/.

9. Haw-Jyh Chiu, Donald A. Fischman, and Ulrich Hammerling, "Vitamin A Depletion Causes Oxidative Stress, Mitochondrial Dysfunction, and PARP-1-Dependent Energy Deprivation," *FASEB Journal* 22, no. 11 (2008): 3878–87, https://www.ncbi.nlm.nih.gov/pmc/articles/PMC2574026/.

10. Maurice Halder et al., "Vitamin K: Double Bonds Beyond Coagulation Insights into Differences Between Vitamin K1 and K2 in Health and Disease," *International Journal of Molecular Sciences* 20, no. 4 (2019): 896, https:// www.ncbi.nlm.nih.gov/pmc/articles/PMC6413124/.

11. Chris Masterjohn, "The Ultimate Vitamin K2 Resource," Chris Masterjohn, PhD (blog), December 9, 2016, https://chrismasterjohnphd.com/blog /2016/12/09/the-ultimate-vitamin-k2-resource/.

12. "Americans Do Not Get All the Nutrients They Need from Food," Council for Responsible Nutrition, https://www.crnusa.org/resources/americans-do -not-get-all-nutrients-they-need-food.

13. Saliha Rizvi et al., "The Role of Vitamin E in Human Health and Some Diseases," *Sultan Qaboos University Medical Journal* 14, no. 2 (2014): e157–65, https://www.ncbi.nlm.nih.gov/pmc/articles/PMC3997530/.

14. Giorgio La Fata, Peter Weber, and M. Hasan Mohajeri, "Effects of Vitamin E on Cognitive Performance During Ageing and in Alzheimer's Disease," *Nutrients* 6, no. 12 (2014): 5453–72, https://www.ncbi.nlm.nih.gov/pmc/articles /PMC4276978/.

15. Dave Asprey, "The Top 7 Anti-inflammatory Herbs and Spices for Bulletproof Cooking," Dave Asprey (blog), https://daveasprey.com/best-anti-inflam matory-herbs-and-spices/.

16. Gary W. Small et al., "Memory and Brain Amyloid and Tau Effects of a Bioavailable Form of Curcumin in Non-demented Adults: A Double-Blind, Placebo-Controlled 18-Month Trial," *American Journal of Geriatric Psychiatry* 26, no. 3 (2018): 266–77, https://www.ajgponline.org/article/S1064 -7481(17)30511-0/fulltext.

17. A. Khajuria, N. Thusu, and U. Zutshi, "Piperine Modulates Permeability Characteristics of Intestine by Inducing Alterations in Membrane Dynamics: Influence on Brush Border Membrane Fluidity, Ultrastructure and Enzyme Kinetics," *Phytomedicine* 9, no. 3 (2002): 224–31, https://pubmed.ncbi.nlm .nih.gov/12046863/.

18. Jennifer M. Ellis and Prabashni Reddy, "Effects of *Panax Ginseng* on Quality of Life," *Annals of Pharmacotherapy* 36, no. 3 (2002): 375–79, https://pubmed.ncbi.nlm.nih.gov/11895046/.

19. Jonathon L. Reay, Andrew B. Scholey, and David O. Kennedy, "Panax ginseng (G115) Improves Aspects of Working Memory Performance and Subjective Ratings of Calmness in Healthy Young Adults," *Human Psychopharmacology* 25, no. 6 (2010): 462–71, https://pubmed.ncbi.nlm.nih.gov/20737519/.

20. K. Asano et al., "Effect of Eleutherococcus senticosus Extract on Human Physical Working Capacity," *Planta Medica* 3 (1986): 175–77, https://pubmed.ncbi.nlm.nih.gov/3749339/.

21. A. F. G. Cicero et al., "Effects of Siberian Ginseng (Eleutherococcus senticosus maxim.) on Elderly Quality of Life: A Randomized Clinical Trial," *Archives of Gerontology and Geriatrics* 38, suppl. (2004): 69–73, https://pubmed.ncbi.nlm.nih.gov/15207399/; Jip Kuo et al., "The Effect of Eight Weeks of Supplementation with Eleutherococcus senticosus on Endurance Capacity and Metabolism in Human," *Chinese Journal of Physiology* 53, no. 2 (2010): 105–11, https://pubmed.ncbi.nlm.nih.gov/21793317/.

22. J. Szolomicki et al., "The Influence of Active Components of *Eleutherococcus senticosus* on Cellular Defence and Physical Fitness in Man," *Phytotherapy Research* 14, no. 1 (2000): 30–35, https://pubmed.ncbi.nlm.nih.gov/10641044.

23. Priyanga Ranasinghe et al., "Medicinal Properties of 'True' Cinnamon (*Cinnamomum zeylanicum*): A Systematic Review," *BMC Complementary Medicine and Therapies* 13 (2013): art. 275, https://www.ncbi.nlm.nih.gov/pmc/articles/PMC3854496/.

24. "Cinnamon," Memorial Sloan Kettering Cancer Center, June 8, 2021, https://www.mskcc.org/cancer-care/integrative-medicine/herbs/cinnamon.

25. Toby Lawrence, "The Nuclear Factor NF-κB Pathway in Inflammation," *Cold Spring Harbor Perspectives in Biology* 1, no. 6 (2009): a001651, https://www.ncbi.nlm.nih.gov/pmc/articles/PMC2882124/.

26. Ranasinghe et al., "Medicinal Properties of 'True' Cinnamon (*Cinnamomum zeylanicum*)."

27. "Cassia Cinnamon with High Coumarin Contents to Be Consumed in Moderation," German Federal Institute for Risk Assessment (BfR), September 2012, https://www.bfr.bund.de/en/press_information/2012/26/cassia_cinnamon_with_high_coumarin_contents_to_be_consumed_in_moderation-131836.html.

28. Somaye Ardebili Dorri et al., "Involvement of Brain-Derived Neurotrophic Factor (BDNF) on Malathion Induced Depressive-like Behavior in Subacute Exposure and Protective Effects of Crocin," *Iranian Journal of Basic Medical Sciences* 18, no. 10 (2015): 958–66, https://www.ncbi.nlm.nih.gov/pmc/articles/PMC4686579/.

29. Mohammad Reza Khazdair et al., "The Effects of *Crocus sativus* (Saffron) and Its Constituents on Nervous System: A Review," *Avicenna Journal of Phytomedicine* 5, no. 5 (2015): 376–91, https://www.ncbi.nlm.nih.gov/pmc/articles/PMC4599112/.

30. Kingshuk Lahon and Swarnamoni Das, "Hepatoprotective Activity of Ocimum sanctum Alcoholic Leaf Extract Against Paracetamol-Induced Liver

Damage in Albino Rats," *Pharmacognosy Research* 3, no. 1 (2011): 13–18, https://pubmed.ncbi.nlm.nih.gov/21731390/.

31. K. P. Bhargava and N. Singh, "Anti-stress Activity of Ocimum sanctum Linn," *Indian Journal of Medical Research* 73 (1981): 443–51, https://pubmed.ncbi.nlm.nih.gov/7275241/.

32. M. Abidov et al., "Extract of *Rhodiola rosea* Radix Reduces the Level of C-Reactive Protein and Creatinine Kinase in the Blood," *Bulletin of Experimental Biology and Medicine* 138, no. 1 (2004): 63–64, https://pubmed.ncbi.nlm.nih.gov/15514725/.

33. V. Darbinyan et al., "*Rhodiola rosea* in Stress Induced Fatigue—A Double Blind Cross-over Study of a Standardized Extract SHR-5 with a Repeated Low-Dose Regimen on the Mental Performance of Healthy Physicians During Night Duty," *Phytomedicine* 7, no. 5 (2000): 365–71, https://pubmed.ncbi.nlm.nih.gov/11081987/.

34. V. Darbinyan et al., "Clinical Trial of *Rhodiola rosea* L. Extract SHR-5 in the Treatment of Mild to Moderate Depression," *Nordic Journal of Psychiatry* 61, no. 5 (2007): 343–48, https://pubmed.ncbi.nlm.nih.gov/17990195/.

35. Barry S. Oken, "Effects of Sage on Memory and Mental Performance in Alzheimer's Disease Patients," ClinicalTrials.gov, October 29, 2014, https://clinicaltrials.gov/ct2/show/NCT00110552.

36. Jueun Oh et al., "Syk/Src Pathway–Targeted Inhibition of Skin Inflammatory Responses by Carosic Acid," *Mediators of Inflammation* 2012, no. 1 (2012): 781375, https://pubmed.ncbi.nlm.nih.gov/22577255/.

37. Magali Chohan, Declan P. Naughton, and Elizabeth I. Opara, "Determination of Superoxide Dismutase Mimetic Activity in Common Culinary Herbs," *SpringerPlus* 3 (2014): 578, https://pubmed.ncbi.nlm.nih.gov/25332878/.

38. Matthew P. Pase et al., "The Cognitive-Enhancing Effects of Bacopa monnieri: A Systematic Review of Randomized, Controlled Human Clinical Trials," *Journal of Alternative and Complementary Medicine* 18, no. 7 (2012): 647–52, https://pubmed.ncbi.nlm.nih.gov/22747190/.

39. Carlo Calabrese et al., "Effects of a Standardized *Bacopa monnieri* Extract on Cognitive Performance, Anxiety, and Depression in the Elderly: A Randomized, Double-Blind, Placebo-Controlled Trial," *Journal of Alternative and Complementary Medicine* 14, no. 6 (2008): 707–13, https://www.ncbi.nlm.nih.gov/pmc/articles/PMC3153866/.

40. Kieran Cooley et al., "Naturopathic Care for Anxiety: A Randomized Controlled Trial ISRCTN78958974," *PLoS ONE* 4, no. 8 (2009): e6628, https://pubmed.ncbi.nlm.nih.gov/19718255/.

41. K. Chandrasekhar, Jyoti Kapoor, and Sridhar Anishetty, "A Prospective, Randomized Double-Blind, Placebo-Controlled Study of Safety and Efficacy of a High-Concentration Full-Spectrum Extract of Ashwagandha Root in Reducing Stress and Anxiety in Adults," *Psychological Medicine* 34, no. 3 (2012): 255–62, https://pubmed.ncbi.nlm.nih.gov/23439798/.

42. Abhinav Grover et al., "Computational Evidence to Inhibition of Human Acetyl Cholinesterase by Withanolide A for Alzheimer Treatment," *Journal of Biomolecular Structure and Dynamics* 29, no. 4 (2012): 651–62, https://pubmed.ncbi.nlm.nih.gov/22208270/.

43. Ibid.; Md Ejaz Ahmed et al., "Attenuation of Oxidative Damage–Associated

Cognitive Decline by *Withania somnifera* in Rat Model of Streptozotocin-Induced Cognitive Impairment," *Protoplasma* 250, no. 5 (2013): 1067–78, https://pubmed.ncbi.nlm.nih.gov/23340606/.

44. Serena Coppola et al. "Potential Clinical Applications of the Postbioitic Butyrate in Human Skin Diseases," *Molecules* 27, no. 6 (2022): 1849, https://www.ncbi.nlm.nih.gov/pmc/articles/PMC8949901/.

45. Megan W. Bourassa et al., "Butyrate, Neuroepigenetics and the Gut Microbiome: Can a High Fiber Diet Improve Brain Health?," *Neuroscience Letters* 625 (2016): 56–63, https://www.ncbi.nlm.nih.gov/pmc/articles/PMC4903954/.

46. A. Singh et al., "Orally Administered Urolithin A Is Safe and Modulates Muscle and Mitochondrial Biomarkers in Elderly," *Innovation in Aging* 1, suppl. 1 (2017): 1223–24, https://www.ncbi.nlm.nih.gov/pmc/articles/PMC6183836/.

47. K. A. Bauerly et al., "Pyrroloquinoline Quinone Nutritional Status Alters Lysine Metabolism and Modulates Mitochondrial DNA Content in the Mouse and Rat," *Biochimica et Biophysica Acta* 1760, no. 11 (2006): 1741–48, https://pubmed.ncbi.nlm.nih.gov/17029795/.

48. Calliandra B. Harris et al., "Dietary Pyrroloquinoline Quinone (PQQ) Alters Indicators of Inflammation and Mitochondrial-Related Metabolism in Human Subjects," *Journal of Nutritional Biochemistry* 24, no. 12 (2013): 2076–84, https://www.sciencedirect.com/science/article/pii/S0955286313001599.

49. B.-Q. Zhu et al, "Pyrroloquinoline Quinone (PQQ) Decreases Myocardial Infarct Size and Improves Cardiac Function in Rat Models of Ischemia and Ischemia/Reperfusion," *Cardiovascular Drugs and Therapy* 18, no. 6 (2004): 421–31, https://pubmed.ncbi.nlm.nih.gov/15770429/.

50. Brittany Sood and Michael Keenaghan, "Coenzyme Q10," National Library of Medicine, National Center for Biotechnology Information, January 19, 2022, https://www.ncbi.nlm.nih.gov/books/NBK531491/.

51. Heather M. Wilkins et al., "Oxaloacetate Activates Brain Mitochondrial Biogenesis, Enhances the Insulin Pathway, Reduces Inflammation and Stimulates Neurogenesis," *Human Molecular Genetics* 23, no. 24 (2014): 6528–41, https://www.ncbi.nlm.nih.gov/pmc/articles/PMC4271074/.

52. Santica M. Marcovina et al., "Translating the Basic Knowledge of Mitochondrial Functions to Metabolic Therapy: Role of L-Carnitine," *Translational Research* 161, no. 2 (2013): 73–84, https://www.ncbi.nlm.nih.gov/pmc/articles/PMC3590819/.

53. Michael J. Lopez and Shamim S. Mohiuddin, "Biochemistry, Essential Amino Acids," National Library of Medicine, National Center for Biotechnology Information, March 18, 2022, https://www.ncbi.nlm.nih.gov/books/NBK557845/.

54. Stefan M. Pasiakos et al., "Leucine-Enriched Essential Amino Acid Supplementation During Moderate Steady State Exercise Enhances Postexercise Muscle Protein Synthesis," *American Journal of Clinical Nutrition* 94, no. 3 (2011): 809–18, https://pubmed.ncbi.nlm.nih.gov/21775557/.

55. M. Lucà-Moretti, "A Comparative, Double-Blind, Triple Crossover Net Nitrogen Utilization Study Confirms the Discovery of the Master Amino Acid Pattern," *Annals of the Royal National Academy of Medicine of Spain* 115,

no. 2 (1998), https://www.puriumcorporate.com/purium1/php_uploads/Studies/mac_comparative_study.pdf.

56. E. Proksch et al., "Oral Intake of Specific Bioactive Collagen Peptides Reduces Skin Wrinkles and Increases Dermal Matrix Synthesis," *Skin Pharmacology and Physiology* 27 (2014): 113–19, https://www.karger.com/Article/Abstract/355523.

57. M. Schunck and S. Oesser, "Specific Collagen Peptides Benefit the Biosynthesis of Matrix Molecules of Tendons and Ligaments," *Journal of the International Society of Sports Nutrition* 10, suppl. 1 (2013): 23, https://www.ncbi.nlm.nih.gov/pmc/articles/PMC4045593/.

58. Daniel König et al., "Specific Collagen Peptides Improve Bone Mineral Density and Bone Markers in Postmenopausal Women—A Randomized Controlled Study," *Nutrients* 10, no. 1 (2018): 97, https://www.ncbi.nlm.nih.gov/pmc/articles/PMC5793325/.

5: CHARGE UP WITH MINERALS

1. "Phytase and Phytate Degradation in Humans," *Nutrition Reviews* 47, no. 5 (1989):155–57, https://pubmed.ncbi.nlm.nih.gov/2541385.

2. Seung-Kwon Myung et al., "Calcium Supplements and Risk of Cardiovascular Disease: A Meta-analysis of Clinical Trials," *Nutrients* 13, no. 2 (2021): 368, https://www.ncbi.nlm.nih.gov/pmc/articles/PMC7910980.

3. Forrest H. Nielsen, "Ultratrace Minerals," USDA Agriculture Research Service, January 1999, https://www.researchgate.net/publication/48855095_Ultratrace_minerals.

4. Asadi Shahmirzadi et al., "Alpha-Ketoglutarate, an Endogenous Metabolite, Extends Lifespan and Compresses Morbidity in Aging Mice," *Cell Metabolism* 32, no. 3 (2020): 447–56, https://doi.org/10.1016/j.cmet.2020.08.004; Randall M. Chin et al., "The Metabolite A-Ketoglutarate Extends Lifespan by Inhibiting ATP Synthase and TOR," *Nature* 510 (2014): 397–40, https://doi.org/10.1038/nature13264; Nan Wu et al., "Alpha-Ketoglutarate: Physiological Functions and Applications," *Biomolecules and Therapeutics* 24, no. 1 (2016): 1–8, https://doi.org/10.4062/biomolther.2015.078; T. Niemiec et al., "Alpha-Ketoglutarate Stabilizes Redox Homeostasis and Improves Arterial Elasticity in Aged Mice," *Journal of Physiology and Pharmacology* 62, no. 1 (2011): 37–43, https://pubmed.ncbi.nlm.nih.gov/21451208/.

5. Ward Dean and Jim English, "Calcium AEP: Membrane Integrity Factor Aids Treatment of Multiple Sclerosis, Asthma and Osteoporosis," Nutrition Review, April 19, 2013, https://nutritionreview.org/2013/04/calcium-aep-membrane-integrity-factor-aids-treatment-multiple-sclerosis-asthma-osteoporosis/.

6. Jakub Chycki et al., "Chronic Ingestion of Sodium and Potassium Bicarbonate, with Potassium, Magnesium and Calcium Citrate Improves Anaerobic Performance in Elite Soccer Players," *Nutrients* 10, no. 11 (2018): 1610, https://www.ncbi.nlm.nih.gov/pmc/articles/PMC6266022/.

7. Suzy V. Torti and Frank M. Torti, "Iron and Cancer: More Ore to Be Mined," *Nature Reviews Cancer* 13, no. 5 (2013): 342–55, https://www.ncbi.nlm.nih.gov/pmc/articles/PMC4036554/.

6: PICK YOUR TARGET

1. James G. Ferry and Christopher H. House, "The Stepwise Evolution of Early Life Driven by Energy Conservation," *Molecular Biology and Evolution* 23, no. 6 (2006): 1286–92, https://academic.oup.com/mbe/article/23/6/1286 /1055368?login=false; Geoffrey M. Cooper, *The Cell: A Molecular Approach*, 2nd ed. (Sunderland, MA: Sinauer Associates, 2000), https://www.ncbi.nlm .nih.gov/books/NBK9903/.

7: HACK TARGET: STRENGTH AND CARDIOVASCULAR FITNESS

1. "Running Injuries," Yale Medicine, https://www.yalemedicine.org/conditions /running-injury.
2. Markus MacGill, "What Is a Normal Blood Pressure Reading?," *Medical News Today*, March 20, 2022, https://www.medicalnewstoday.com/articles/270644.
3. John Jaquish, interviewed by Dave Asprey, "Exercise Biohacks to Induce Bone Density and Grow Muscles in Less Time," *Bulletproof* podcast no. 427, https://daveasprey.com/exercise-biohacks-bone-density-grow-muscles-427/.
4. Richard S. Metcalfe et al., "Towards the Minimal Amount of Exercise for Improving Metabolic Health: Beneficial Effects of Reduced-Exertion High-Intensity Interval Training," *European Journal of Applied Physiology* 112, no. 7 (2012): 2767–75, https://pubmed.ncbi.nlm.nih.gov/22124524/.
5. José S. Ruffino et al., "A Comparison of the Health Benefits of Reduced-Exertion High-Intensity Interval Training (REHIT) and Moderate-Intensity Walking in Type 2 Diabetes Patients," *Applied Physiology, Nutrition, and Metabolism* 42, no. 2 (2017): 202–08, https://core.ac.uk/reader/77612359?utm _source=linkout.
6. Lance Dalleck, interviewed by Dave Asprey, "Fast Fitness! 40 Seconds, 3 Times a Week," *The Human Upgrade™ with Dave Asprey* podcast no. 657, January 7, 2020, https://radiopublic.com/the-human-upgrade-with-dave-aspre -WP5096/s1!171bb.
7. Tom F. Cuddy, Joyce S. Ramos, and Lance C. Dalleck, "Reduced Exertion High-Intensity Interval Training Is More Effective at Improving Cardiorespiratory Fitness and Cardiometabolic Health than Traditional Moderate-Intensity Continuous Training," *International Journal of Environmental Research and Public Health* 16, no. 3 (2019): 483, https://www.ncbi.nlm.nih.gov/pmc /articles/PMC6388288/#B11-ijerph-16-00483.
8. Valentín E. Fernández-Elías et al., "Relationship Between Muscle Water and Glycogen Recovery After Prolonged Exercise in the Heat in Humans," *European Journal of Applied Physiology* 115, no. 9 (2015): 1919–26, https:// pubmed.ncbi.nlm.nih.gov/25911631/.
9. Jonathan P. Little et al., "An Acute Bout of High-Intensity Interval Training Increases the Nuclear Abundance of PCG-1α and Activates Mitochondrial Biogenesis in Human Skeletal Muscle," *American Journal of Physiology* 300, no. 6 (2021): R1303–10, https://journals.physiology.org/doi/full/10.1152/ajp regu.00538.2010?rfr_dat=cr_pub++0pubmed&url_ver=Z39.88-2003&rfr _id=ori%3Arid%3Acrossref.org.
10. Ruffino et al., "A Comparison of the Health Benefits of Reduced-Exertion High-Intensity Interval Training (REHIT)."

11. D. M. Morris, J. T. Kearney, and E. R. Burke, "The Effects of Breathing Supplemental Oxygen During Altitude Training on Cycling Performance," *Journal of Science and Medicine in Sport* 3, no. 2 (2000): 165–75, https://www.jsams.org/article/S1440-2440(00)80078-X/pdf.

12. Cuddy, Ramos, and Dalleck, "Reduced Exertion High-Intensity Interval Training Is More Effective."

13. Gretl Lam et al., "Hyperbaric Oxygen Therapy: Exploring the Clinical Evidence," *Advances in Skin & Wound Care* 30, no. 4 (2017): 181–90, https://journals.lww.com/aswcjournal/Fulltext/2017/04000/Hyperbaric_Oxygen_Therapy__Exploring_the_Clinical.8.aspx.

14. Stephen R. Thom, "Hyperbaric Oxygen: Its Mechanisms and Efficacy," *Plastic and Reconstructive Surgery* 127, suppl. 1 (2011): 131S–141S, https://www.ncbi.nlm.nih.gov/pmc/articles/PMC3058327/.

15. Junichi Suzuki, "Endurance Performance Is Enhanced by Intermittent Hyperbaric Exposure via Up-regulation of Proteins Involved in Mitochondrial Biogenesis in Mice," *Physiological Reports* 5, no. 15 (2017): e13349, https://physoc.onlinelibrary.wiley.com/doi/full/10.14814/phy2.13349.

16. Amir Hadanny et al., "Effects of Hyperbaric Oxygen Therapy on Mitochondrial Respiration and Physical Performance in Middle-Aged Athletes: A Blinded, Randomized Controlled Trial," *Sports Medicine: Open* 8, no. 22 (2022), https://sportsmedicine-open.springeropen.com/articles/10.1186/s40798-021-00403-w.

8: HACK TARGET: ENERGY LEVEL AND METABOLISM

1. Qian Cheng and Meng-Lu Qian, "Piezoelectric Effect of Cell's Membrane," *Journal of the Acoustical Society of America* 131, no. 4 (2012): 3246, https://asa.scitation.org/doi/10.1121/1.4708114.

2. Xingxing Shi et al., "Ultrasound-Activable Piezoelectric Membranes for Accelerating Wound Healing," *Biomaterials Science* 10, no. 3 (2022): 692–701, https://pubmed.ncbi.nlm.nih.gov/34919105/; D. Denning et al., "Piezoelectric Properties of Aligned Collagen Membranes," *Journal of Biomedical Materials Research Part B: Applied Biomaterials* 102, no. 2 (2014): 284–92, https://pubmed.ncbi.nlm.nih.gov/24030958/.

3. William R. Thompson, Sherwin S. Yen, and Janet Rubin, "Vibration Therapy: Clinical Applications in Bone," *Current Opinion in Endocrinology, Diabetes Obesity* 21, no. 6 (2014): 447–53, https://www.ncbi.nlm.nih.gov/pmc/articles/PMC4458848/; "Update on Vibration Therapy for Bone Health," Harvard Health Publishing, October 1, 2011, https://www.health.harvard.edu/womens-health/update-on-vibration-therapy-for-bone-health.

4. Angela Navarrete-Opazo and Gordon S. Mitchell, "Therapeutic Potential of Intermittent Hypoxia: A Matter of Dose," *American Journal of Physiology: Regulatory, Integrative and Comparative Physiology* 307, no. 10 (2014): R1181–97, https://pubmed.ncbi.nlm.nih.gov/25231353/; E. A. Dale, F. Ben Mabrouk, and G. S. Mitchell, "Unexpected Benefits of Intermittent Hypoxia: Enhanced Respiratory and Nonrespiratory Motor Function," *Physiology* 29, no. 1 (2014): 39–48, https://www.ncbi.nlm.nih.gov/pmc/articles/PMC4073945/.

5. Tatiana V. Serebrovskaya, "Intermittent Hypoxia Training as Non-pharmacologic Therapy for Cardiovascular Diseases: Practical Analysis on Methods and Equipment," *Experimental Biology and Medicine* 241, no. 15 (2016): 17087–23, https://pubmed.ncbi.nlm.nih.gov/27407098/.

6. Dave Asprey, "Deep Breathing Strengthens Your Brain and Boosts Attention Span, Says New Study," Dave Asprey (blog), https://daveasprey.com/breathing-sharpens-brain-study/.

7. Sébastien Herzig and Reuben J. Shaw, "AMPK: Guardian of Metabolism and Mitochondrial Homeostasis," *Nature Reviews Molecular Cell Biology* 19, no. 2 (2018): 121–35, https://www.ncbi.nlm.nih.gov/pmc/articles/PMC5780224/.

8. Mauricio and Raul Uranga, "Breath Control, Exercises & Sets: Hypoxic Training," Skills NT, June 11, 2020, https://skillswimming.com/breath-control-swimming/.

9. "Genetics of Oxygen Deprivation in Marine Mammals and Humans," Duke University, https://bassconnections.duke.edu/virtual-showcase/genetics-oxygen-deprivation-marine-mammals-and-humans.

10. Wim Hof, "This Is 'Iceman' Wim Hof!," YouTube, April 19, 2019, https://www.youtube.com/watch?v=MgKdHG6MQ0g.

11. Sukanya Suresh, Praveen Kumar Rajvanshi, and Constance T. Noguchi, "The Many Facets of Erythropoietin Physiologic and Metabolic Response," *Frontiers in Physiology* 10 (2019): 1534, https://www.frontiersin.org/articles/10.3389/fphys.2019.01534/full.

12. Rashi Singhal and Yatrik M. Shah, "Oxygen Battle in the Gut: Hypoxia and Hypoxia-Inducible Factors in Metabolic and Inflammatory Responses in the Intestine," *Journal of Biological Chemistry* 295, no. 30 (2020): 10493–505, https://www.ncbi.nlm.nih.gov/pmc/articles/PMC7383395/.

13. Zachary Long, "The Science of Blood Flow Restriction," The Barbell Physio, https://thebarbellphysio.com/science-blood-flow-restriction-training/.

14. Vahid Fekri-Kurabbaslou, Sara Shams, and Sadegh Amani-Shalamzari, "Effect of Different Recovery Modes During Resistance Training with Blood Flow Restriction on Hormonal Levels and Performance in Young Men: A Randomized Controlled Trial," *BMC Sports Science, Medicine and Rehabilitation* 14, art. 47 (2022), https://bmcsportsscimedrehabil.biomedcentral.com/articles/10.1186/s13102-022-00442-0.

15. Stephen John Pearson and Syed Robiul Hussain, "A Review on the Mechanisms of Blood-Flow Restriction Resistance Training–Induced Muscle Hypertrophy," *Sports Medicine* 45, no. 2 (2015): 187–200, https://pubmed.ncbi.nlm.nih.gov/25249278/.

16. Jim Stray-Gundersen, interviewed by Dave Asprey, "How Blood Flow Restriction Can Revolutionize Your Fitness," *Bulletproof* podcast no. 705, https://daveasprey.com/jim-stray-gundersen-705/.

17. Paul R. T. Kuzyk and Emil H. Schemitsch, "The Science of Electrical Stimulation Therapy for Fracture Healing," *Indian Journal of Orthopaedics* 43, no. 2 (2009): 127–31, https://www.ncbi.nlm.nih.gov/pmc/articles/PMC2762253/.

18. Fernanda Martini et al., "Bone Morphogenetic Protein-2 Signaling in the Osteogenic Differentiation of Human Bone Marrow Mesenchymal Stem Cells Induced by Pulsed Electromagnetic Fields," *International Journal of Mo-

lecular Sciences 21, no. 6 (2020): 2104, https://www.ncbi.nlm.nih.gov/pmc /articles/PMC7139765/.

19. Beatrice Borges, Ronald Hosek, and Susan Esposito, "Effects of PEMF (Pulsed Electromagnetic Field) Stimulation on Chronic Pain and Anxiety Utilizing Decreased Treatment Frequency and Duration Application," conference abstract, International Symposium on Clinical Neuroscience, May 24–26, 2019, Orlando, Florida, https://www.frontiersin.org/10.3389%2Fconf.fneur.2019 .62.00007/event_abstract.

9: HACK TARGET: BRAIN AND NEURO-FITNESS

1. Christel Kannegiesser-Leitner and Ralphe Warnke, "Hemoencephalography: HEG Based Neurofeedback Practically Introduced as a Smart and Easy-to-Use Training Method in ADD/ADHD, Dyslexia and Other Learning Disorders," *Applied Psychophysiology and Biofeedback* 40, no. 4 (2015): 364–65, https://www .researchgate.net/publication/290045172_Hemoencephalography_HEG _Based_Neurofeedback_Practically_Introduced_as_a_Smart_and_Easy-to -Use_Training_Method_in_ADDADHD_Dyslexia_and_Other_Learning _Disorders.

2. Mireia Serra-Sala, Carme Timoneda-Gallart, and Frederic Pérez-Álvarez, "Clinical Usefulness of Hemoencephalography Beyond the Neurofeedback," *Neuropsychiatric Disease and Treatment* 2016, no. 12 (2016): 1173–80, https://www.ncbi.nlm.nih.gov/pmc/articles/PMC4869785/.

3. Cleveland Clinic, "Temporomandibular Joint (TMJ) Disorders," Cleveland Clinic, June 21, 2021, https://my.clevelandclinic.org/health/diseases/15066 -temporomandibular-disorders-tmd-overview.

4. Pedro Shiozawa et al., "Transcutaneous Vagus and Trigeminal Nerve Stimulation for Neuropsychiatric Disorders: A Systematic Review," *Arquivos de Neuro-psiquiatria* 72, no. 7 (2014): 542–47, https://pubmed.ncbi.nlm.nih.gov /25054988/.

5. Yoko Yamazaki et al., "Modulation of Paratrigeminal Nociceptive Neurons Following Temporomandibular Joint Inflammation in Rats," *Experimental Neurology* 214, no. 2 (2008): 209–18, https://pubmed.ncbi.nlm.nih.gov/187 78706/.

6. Laurie Kelly McCorry, "Physiology of the Autonomic Nervous System," *American Journal of Pharmaceutical Education* 71, no. 4 (2007): 78, https:// www.ncbi.nlm.nih.gov/pmc/articles/PMC1959222/.

7. Jordan Fallis, "How to Stimulate Your Vagus Nerve for Better Mental Health," University of Ottawa, January 21, 2017, https://sass.uottawa.ca /sites/sass.uottawa.ca/files/how_to_stimulate_your_vagus_nerve_for_better _mental_health_1.pdf.

8. "The Safe and Sound Protocol (SSP)," Unyte and Integrated Listening Systems, https://integratedlistening.com/ssp-safe-sound-protocol.

9. Russel Lazarus, "How Is a Brock String Used?," Optometrists Network, November 4, 2021, https://www.optometrists.org/vision-therapy/guide-to -vision-therapy/vision-therapy-faqs/how-is-a-brock-string-used/.

10. Harpreet Shinmar et al., "Weeklong Improved Colour Contrasts Sensitivity After Single 670 nm Exposures Associated with Enhanced Mitochondrial

Function," *Scientific Reports* 11 (2021): art. 22872, https://www.nature.com/articles/s41598-021-02311-1.

11. Ibid.

12. Christine Blume, Corrado Barbazza, and Manuel Spitschan, "Effects of Light on Human Circadian Rhythms, Sleep and Mood," *Somnologie* 23, no. 3 (2019): 147–56, https://www.ncbi.nlm.nih.gov/pmc/articles/PMC6751071/.

13. "Melatonin: What You Need to Know," National Center for Complementary and Integrative Health, July 2022, https://www.nccih.nih.gov/health/melatonin-what-you-need-to-know.

10: HACK TARGET: RESILIENCE AND RECOVERY

1. Alison Moodie, "How to Sleep Better: Science-Backed Sleep Hacks to Wake Up Ready to Go," Bulletproof, June 15, 2021, https://www.bulletproof.com/sleep/sleep-hacks/how-to-sleep-better/.

2. Margeaux M. Schade et al., "Enhancing Slow Oscillations and Increasing N3 Sleep Proportion with Supervised, Non-Phase-Locked Ping Noise and Other Non-standard Auditory Stimulation During NREM Sleep," *Nature and Science of Sleep* 12 (2020): 411–29, https://www.ncbi.nlm.nih.gov/pmc/articles/PMC7364346/.

3. Fred Grover, Jr., Jon Weston, and Michael Weston, "Acute Effects of Near Infrared Light Therapy on Brain State in Healthy Subjects as Quantified by EEG Measures," *Photomedicine and Laser Surgery* 35, no. 3 (2017): 136–41, https://pubmed.ncbi.nlm.nih.gov/27855264/.

4. Natalya A. Zhevago and Kira A Samilova, "Pro- and Anti-inflammatory Cytokine Content in Human Peripheral Blood After Its Transcutaneous (in Vivo) and Direct (in Vitro) Irradiation with Polychromatic Visible and Infrared Light," *Photomedicine and Laser Surgery* 24, no. 2 (2006): 129–39, https://pubmed.ncbi.nlm.nih.gov/16706691/.

5. Pinar Avci et al., "Low-Level Laser (Light) Therapy (LLLT) in Skin: Stimulating, Healing, Restoring," *Seminars in Cutaneous Medicine and Surgery* 32, no. 1 (2013): 41–52, https://www.ncbi.nlm.nih.gov/pmc/articles/PMC4126803/.

6. Roma Parikh et al., "Skin Exposure to UVB Light Induces a Skin-Brain-Gonad Axis and Sexual Behavior," *Cell Reports* 36, no. 8 (2021): 109579, https://www.cell.com/cell-reports/fulltext/S2211-1247(21)01013-5?_return URL=https%3A%2F%2Flinkinghub.elsevier.com%2Fretrieve%2Fpii%2FS2211124721010135%3Fshowall%3Dtrue.

7. Nayan Huang et al., "Safety and Efficacy of 630-nm Red Light on Cognitive Function in Older Adults with Mild to Moderate Alzheimer's Disease: Protocol for a Randomized Controlled Study," *Frontiers in Aging Neuroscience* 12 (2020): art. 143, https://www.frontiersin.org/articles/10.3389/fnagi.2020.00143/full.

8. Roberta Chow et al., "Guidelines Versus Evidence: What We Can Learn from the Australian Guideline for Low-Level Laser Therapy in Knee Osteoarthritis? A Narrative Review," *Lasers in Medical Science* 36 (2021): 249–58, https://link.springer.com/article/10.1007/s10103-020-03112-0; Marco Maiello et al., "Infrared Light for Generalized Anxiety Disorder: A Pilot Study,"

Photobiomodulation, Photomedicine, and Laser Surgery 37, no. 10 (2019), https://www.liebertpub.com/doi/10.1089/photob.2019.4677.

9. Alexander Panossian and Georg Wikman, "Effects of Adaptogens on the Central Nervous System and the Molecular Mechanisms Associated with Their Stress-Protective Activity," *Pharmaceuticals* 3, no. 1 (2010): 188–224, https://www.ncbi.nlm.nih.gov/pmc/articles/PMC3991026/.

10. Andrew D. Huberman, "Supercharge Exercise Performance & Recovery with Cooling," Huberman Lab, May 10, 2021, https://hubermanlab.com/super charge-exercise-performance-and-recovery-with-cooling/.

11. Nana Chung, Jonghoon Park, and Kiwon Lim, "The Effects of Exercise and Cold Exposure on Mitochondrial Biogenesis I Skeletal Muscle and White Adipose Tissue," *Journal of Exercise Nutrition & Biochemistry* 21, no. 2 (2017): 39–47, https://www.ncbi.nlm.nih.gov/pmc/articles/PMC5545200/.

12. Kathleen A. O'Hagan et al., "PGC-1α Is Coupled to HIF-1α-Dependent Gene Expression by Increasing Mitochondrial Oxygen Consumption," *Proceedings of the National Academy of Sciences of the United States of America* 107, no. 7 (2009): 2188–93, https://www.pnas.org/doi/full/10.1073/pnas.0808801106.

13. Huafeng Zhang et al., "Mitochondrial Autophagy Is an HIF-1-Dependent Adaptive Metabolic Response to Hypoxia," *Journal of Biological Chemistry* 283, no. 16 (2008): 10892–903, https://www.ncbi.nlm.nih.gov/pmc/articles /PMC2447655/.

14. Mansal Denton, "Mitochondria Health: An Exploration of Temperature and Light Therapy," Neurohacker Collective, February 20, 2019, https://neuro hacker.com/mitochondria-health-an-exploration-of-temperature-and-light -therapy-28ff9793-1e48-42f8-9943-bc59446c52fd.

15. Tanjaniina Laukkanen et al., "Association Between Sauna Bathing and Fatal Cardiovascular and All-Cause Mortality Events," *JAMA Internal Medicine* 175, no. 4 (2015): 542–48, https://jamanetwork.com/journals/jamainternal medicine/fullarticle/2130724; Joy Hussain and Marc Cohen, "Clinical Effects of Regular Dry Sauna Bathing: A Systematic Review," *Evidence-Based Complementary and Alternative Medicine* 2018 (2018): art. 1857413, https:// www.ncbi.nlm.nih.gov/pmc/articles/PMC5941775/.

16. "Sauna Health Benefits: Are Saunas Healthy or Harmful?," Harvard Health Publishing, May 14, 2020, https://www.health.harvard.edu/staying-healthy /saunas-and-your-health.

17. FDA Consumer Updates, "Whole Body Cryotherapy (WBC): A 'Cool' Trend That Lacks Evidence, Poses Risks," U.S. Food and Drug Administration, July 5, 2016, https://www.fda.gov/consumers/consumer-updates/whole-body -cryotherapy-wbc-cool-trend-lacks-evidence-poses-risks.

18. Elahu G. Sustarsic et al., "Cardiolipin Synthesis in Brown and Beige Fat Mitochondria Is Essential for Systemic Energy Homeostasis," *Cell Metabolism* 28, no. 1 (2018): 159–74.e11, https://pubmed.ncbi.nlm.nih.gov/29861389/.

19. Susanna Søberg, "Using Cold to Enhance Your Metabolism," reposted from Andrew Huberman, Facebook, November 22, 2021, https://www.facebook .com/susannasoeberg/posts/repost-from-andrew-huberman-phd-using-cold -to-enhance-your-metabolism-many-peopl/316083460518732/.

20. Jonathan A. Lindquist and Peter R. Mertens, "Cold Shock Proteins: From Cellular Mechanisms to Pathophysiology and Disease," *Cell Communication*

and Signaling 16 (2018): art. 63, https://biosignaling.biomedcentral.com/arti
cles/10.1186/s12964-018-0274-6.

21. Ward Dean and Jim English, "Calcium AEP: Membrane Integrity Factor Aids Treatment of Multiple Sclerosis, Asthma and Osteoporosis," Nutrition Review, April 19, 2013, https://nutritionreview.org/2013/04/calcium-aep-membrane -integrity-factor-aids-treatment-multiple-sclerosis-asthma-osteoporosis/.

22. Stavros Lalas, Vassilis Athanasiadis, and Vassilis Dourtoglou, "Humic and Fulvic Acids as Potentially Toxic Metal Reducing Agents in Water," *CLEAN: Soil Air Water* 46, no. 2 (2018): 1700608, https://www.researchgate.net/pub lication/321694288_Humic_and_Fulvic_Acids_as_Potentially_Toxic_Metal _Reducing_Agents_in_Water.

23. Fabrizio De Paolis and Jussi Kukkonen, "Binding of Organic Pollutants to Humic and Fulvic Acids: Influence of pH and the Structure of Humic Mate rial," *Chemosphere* 34, no. 8 (1997): 1693–704, https://www.sciencedirect .com/science/article/abs/pii/S004565359700026X.

11: SPIRITUAL STRENGTH

1. Stephen W. Porges, "The Polyvagal Theory: New Insights into Adaptive Reactions of the Autonomic Nervous System," *Cleveland Clinic Journal of Medicine* 76, suppl. 2 (2009): S86–90, https://www.ncbi.nlm.nih.gov/pmc /articles/PMC3108032/.

2. Laurie B. Agrimson and Lois B. Taft, "Spiritual Crisis: A Concept Analysis," *Journal of Advanced Nursing* 65, no. 2 (2009): 454–61, https://pubmed.ncbi .nlm.nih.gov/19040691/.

12: THE NEXT-LEVEL UPGRADE

1. Gioacchino Tangari et al., "Mobile Health and Privacy: Cross Sectional Study," *BMJ* 373, no. 1248 (2021), https://www.bmj.com/content/373/bmj.n1 248.

2. Jiuwei Gao et al., "Ultra-robust and Extensible Fibrous Mechanical Sensors for Wearable Smart Healthcare," *Advanced Materials* 34, no. 20 (2022): e2107511, https://onlinelibrary.wiley.com/doi/abs/10.1002/adma.202107511.

3. Alla Katsnelson, "Male Researchers Stress Out Rodents," *Nature*, April 28, 2014, https://www.nature.com/articles/nature.2014.15106.

4. Randy J. Nelson et al., "Time-of-Day as a Critical Biological Variable," *Neuroscience & Biobehavioral Reviews* 127 (2021): 740–46, https://www.science direct.com/science/article/abs/pii/S0149763421002190.

5. "ICS Medical Advisory (ICSMA-19-080-01)," Cybersecurity & Infrastructure Security Agency, April 8, 2021, https://www.cisa.gov/uscert/ics/adviso ries/ICSMA-19-080-01.

6. Eliza Strickland and Mark Harris, "Their Bionic Eyes Are Now Obsolete and Unsupported," *IEEE Spectrum*, February 15, 2022, https://spectrum.ieee.org /bionic-eye-obsolete.

7. Simanto Saha et al., "Progress in Brain Computer Interface: Challenges and Opportunities," *Frontiers in Systems Neuroscience* 15, art. 578875 (2021), https://www.frontiersin.org/articles/10.3389/fnsys.2021.578875/full.

8. "Cybin Partnership Case Study—More Effective Treatments: Understanding Psychedelic Neuro Effects," Kernel, https://www.kernel.com/.

13: YOU DO YOU

1. Keith Rowe, "Test!," BrainMD, May 4, 2022, https://brainmd.com/blog/brain-type-test/.
2. Jeff Hawkins, *A Thousand Brains: A New Theory of Intelligence* (New York: Basic Books, 2021).
3. Neil Strauss, *Emergency: This Book Will Save Your Life* (New York: It Books, 2009).
4. "Dr. Sean Stephenson, 1979–2019," SeanStephenson, https://seanstephenson.com/.

14: EVALUATE, PERSONALIZE, REPEAT

1. "hTMA Science—References," Nutritional Balancing.Org, August 9, 2022, https://nutritionalbalancing.org/center/htma/science/articles/htma-references.php.
2. Jay Mohan et al., "Coronary Artery Calcification," StatPearls, May 2, 2022, https://www.ncbi.nlm.nih.gov/books/NBK519037/.

INDEX

A1 and A2 casein protein, 52–54
acacia gum, 71, 92
acesulfame potassium (ace-K), 58
acetyl-L-carnitine, 56, 94, 156
activated charcoal, 209, 214
Active PQQ (pyrroloquinoline
    quinone), 93–94, 156, 176
active straight-leg raises, 150, 153–54
acupuncture and acupressure points:
    devices that flash LED lights over,
        184
    devices that vibrate over, 160
adaptogens, 90–91, 201–2
    see also ashwagandha; ginseng;
        holy basil; *rhodiola rosea*
additives, artificial, 57–58
adenosine triphosphate (ATP), 24, 25,
    93–94, 101, 199, 200
aflatoxins, 54–55
aging, 26, 50, 62, 63, 93, 102, 110,
    240, 242, 269
agricultural practices, 75–76, 98
alcohol, 11, 47, 55, 103
alkalosis, 106
allergies, 78, 107, 245
    histamines and, 45–47
aluminum, 112
Alzheimer's disease, 84, 89, 91, 122,
    239, 240
Amen, Daniel, 151, 192, 252, 268
Amen Clinics, 60
amino acids, 65, 94–95, 131
AMP-activated protein kinase
    (AMPK), 147, 165
anemia, 110
angiogenesis, 164
antibiotics, in agribusinesses, 52, 56

antinutrients, 33–52, 66, 69, 70
    glyphosate, 47–48, 69–70, 76
    lectins, 38, 39–41, 50, 67, 70, 71
    oxalic acid, 38–39, 71, 99
    plants that are low in, 70–71
    see also omega-6 fatty acids; phytic
        acid
antioxidants, 43, 62, 68, 72–73, 84,
    93, 107
    in herbs and spices, 86, 87, 88,
        89, 90
anxiety, 46, 89, 91, 123, 197
apnea, 198
Apollo devices, 160, 180
arterial calcification, 62, 265
artificial additives, 57–58
artificial intelligence (AI), 125,
    243–44, 249, 273
    cardio training and, 120, 148–49
    strength training and, 134, 140–41
ashwagandha, 90–91, 92, 195, 202
aspartame, 57–58
auditory integration therapy (AIT),
    181–82
autolyzed yeast extract, 58
autonomic nervous system, 180
avocados, 43, 70, 71, 106

Bacopa monnieri (water hyssop), 90,
    92, 176
barbells, 132, 134
basil, holy (*Ocimum tenuiflorum*),
    88–89, 91, 195, 202
Baton Rouge device, 188–89
Becker, Robert O., 170
beef, 43
    grass-fed, 49, 67–68, 84, 94, 110

beer can breathing, 150, 152–53
Bell, Mark, 141
Bella, Nikki, 172
beta-carotene, 62, 80
betaine hydrochloride, 100
beta state, 184
Bifidobacterium infantis and *B.
longum*, 92, 93
Big Food, 35, 79, 80, 81, 83, 98–99
biogenic toxins, 54–55
biohacking concept, 12–16, 119
Hacks-Goals matrix, 125–8
next era of, 248–50
bionics, 244–46
birth control pills, 111
black oil plant (*Celastrus
paniculatus*), 176
black pepper, 86, 91
blood clotting, 62–63, 265
blood flow hacks, 167–70
custom blood flow restriction
bands, 167, 169–70
pressure cuffs, 167, 168–69
blood flow in brain, monitoring with
hemoencephalography (HEG),
176, 177–78
blood glucose, 16, 103, 108
blood glucose monitors, 238–39, 266
blood pressure, 78, 103, 105, 108,
137, 180, 206, 209
blue light, 187
bone density, 62
metabolic performance and, 171
osteoporosis and, 35, 62, 78, 82,
97, 107
whole-body vibration and, 157
bone morphogenic protein, 171–72
brain:
development of, 253–54
monitoring of environment and,
175
omega-6 fatty acids and, 41–42
self-awareness and, 175
brain and neuro-fitness, 174–93
EEG evaluation of, 177, 232,
235–36, 237, 268
as target for improvement, 118–19,
120, 121, 122, 123, 126, 129
see also brain and neuro-fitness
training
brain and neuro-fitness training, 126,
174–93
hemoencephalography (HEG)

to monitor blood flow, 176,
177–78
LED brain blinking, 185, 189–90
modulated-light LEDs and lasers,
185, 188–89
neurofeedback, 175–79, 182, 225,
235, 273
reading under red light, 185,
186–87
removing junk light, 185, 187
sound hacks, 181–84
supplements to support, 176
transcranial alternating-current
stimulation (tACS), 191, 192
transcranial direct-current
stimulation (tCDS), 190,
191–92
vagal nerve stimulation, 179–81
brain fog, 15, 26, 45, 46, 63, 92,
107, 122, 189, 204, 252
brain readers, 246–47
BrainTap, 184
brain waves, EEG monitoring of, 177
breathing, 126
bite guard for improvement in,
196, 197–98
controlled, with cold therapy, 203,
208–9
controlled hypoxia, 163–64
functional movement exercises
and, 150, 152–54
hacks for energy level and
metabolism, 163–66
holding your breath, 163, 164
hyperventilating, then exhaling
fully, 164–65
hypoxia induced with hypoxicator,
163, 165–66
for spiritual recovery, 232, 233
tech-guided, 163, 165
Wim Hof, 163, 165, 166, 208, 233
bromelain, 86, 91
brushing face, for lymphatic drainage,
212
Brussels sprouts, 72, 212
B Strong–style belts, 169–70
Buddhist approaches, 6, 223, 258–59
butter, 42, 80, 82, 92–93
grass-fed, 56, 69, 83, 84
butyric acid, or butyrate, 66, 92–93

cabbage, 70, 72, 212
calcium, 35, 38, 48, 63, 72, 73–74,

78, 79, 80, 82, 85, 97, 100–103, 113, 157, 213
  arterial calcification and, 62, 265
  oxalic acid and, 38–39
  ratio of magnesium and, 104
calcium AEP, 102, 103, 113, 213
calcium AKG, 102, 103, 113
calcium carbonate, 100–101
calcium citrate, 101–2
calcium-D-glucarate, 102, 103, 113, 209, 212
calcium fructoborate, 102, 103, 113
camel's milk, 53, 54
camphor, 89–90
cancer, 41, 48, 54, 66, 68, 69, 73, 78, 82, 84, 87, 89, 90, 110, 112, 249
carbohydrates, 51, 66–67, 73
cardiolipin, 41–42, 43, 204, 208
cardio training, 142–50
  AI-guided cycling, 148–49
  high-intensity interval training (HIIT), 120, 142, 145–46
  oxygen hit, 149–50
  reduced exertion high-intensity training (REHIT), 120, 142, 146–48
  steady-state cardio, 142–44
  supplements to prepare for, 131
  tapping power of laziness and, 19–21
  varying-intensity interval training, 142, 144–45
  work-hard philosophy and, 129–33, 142–44
  zone training, 142–44
cardiovascular fitness:
  evaluating your outcome in, 267
  functional movement and, 150–54, 256
  as target for improvement, 118–20, 121, 122, 123, 126, 129
  see also cardio training
carnitine, 68
  acetyl-L-carnitine, 56, 94
carnosic acid, 89
carnosine, 108
carnosol, 89
C-clamp exercise (aka beer can breathing), 150, 152–53
Celastrus paniculatus (black oil plant), 176
cell membranes, 53, 85, 101, 102, 171

chanting, 231
  vocal vibration and, 157, 158
charcoal, activated, 209, 214
chard, 39
cheeses, 47, 112
chicken, 44, 49–51
Chinese traditions, 86, 89, 164
chloride, 35, 100
chocolate, 47, 55, 71, 73, 103, 107
cholesterol, 78, 79, 202
chromium, 35, 36, 265
cinnamon, 86, 87–88, 91
circadian rhythm, 81, 104, 186–87
cobalt, 35, 64
cod liver oil, 79, 80
coenzyme Q10 (CoQ10), 68, 94–95
cofactors, 28, 98, 103
coffee, 55, 71, 73, 92, 131
  Danger Coffee, 64, 96, 113, 214
cognition, improving, 84, 86–87
cold, *see* heat and cold hacks
cold-shock proteins, 210
collagen, 66, 95, 102, 157, 200, 213
colors, artificial, 57–58
compassion, 6, 258
compression garments, 209, 211, 212
control of yourself, 3–6
copper, 35, 36, 48.64, 56, 107–8, 110, 113, 131, 200, 265
cortisol, 269
COVID, 1–6, 45, 107
cow's milk, 52–54
C-reactive protein, or CRP, 89, 91, 102
creatine, 176
cruciferous vegetables, 72, 109
cryotherapy, 203, 207, 210–11
Cunermuspir, 107
curcumin, 86, 91
cycling:
  AI-guided, 148–49
  hypoxia in training for, 163–64
cytochrome c oxidase, 199–200, 201

dairy products, 55, 101
  grass-fed, 53–54, 56, 68–69, 83, 84, 106
  tips on choosing, 52–54
Dalai Lama, 224
Dalleck, Lance, 147
Danger Coffee, 64, 96, 113, 214
depression, 3, 36, 65, 78, 89, 103, 109, 202, 245

detoxification, 66, 194–95, 209–16,
219, 225
activated charcoal and, 214
calcium-D-glucarate and, 102
cleaning up your environment and,
209, 214–15
heat and cold and, 203, 204–5,
206–7
kidney function and, 213
liver function and, 212–13
molybdenum and, 36, 111
raw carrots and, 70
saffron and, 88
diabetes, 36, 73, 78, 103, 112
Diamandis, Peter, 140, 246
diamine oxidase (DAO), 46
didgeridoo, 158–59
digital version of you, 247–48
DNA, 25, 41, 54, 133, 210, 269
DNA methylation tests, 269
dual-energy X-ray absorptiometry
(DEXA), 268–69
duck, 44, 50, 51
dumbbells, 136

electrical hacks, 190–93
electrical muscle stimulation
(EMS), 190, 192–93
transcranial direct-current
stimulation (tCDS), 190, 191–92
electrical muscle stimulation (EMS),
134, 138–40, 190, 192–93
electrical stimulation therapy, 170–71
electroencephalograms (EEGs), 177,
232, 235–36, 237, 268
electrolytes, 102, 106, 131, 209, 213
electromagnetic fields (EMFs),
245–46
electromagnetic hacking, 170–73
pulsed electromagnetic field
(PEMF) therapy, 171–73
eleutero, aka Siberian ginseng, 87,
91, 202
emotion, coming to terms with,
220–23, 224
empathy, 6, 258
endocrine system, 202
energy, hacking the laziness principle
and, 19, 24–30
energy level and metabolism, 155–73
blood flow hacks and, 167–70
breathing hacks and, 163–66
electromagnetic hacks and, 170–73

evaluating your outcome in, 268
glyphosate and, 49
supplements and, 156
as target for improvement, 118–19,
120, 121–22, 123, 126, 129
vibration hacks and, 156–62
energy supplements, 93–94
environment:
cleaning up your own, 209,
214–15
epigenetics and, 242–43
enzymes, 65, 97–98, 103, 107, 206
epigenetics, 242–43, 269
equanimity, 6, 258–59
estrogen, 62, 212, 269
evaluating your outcomes, 263–70
continuous improvement and, 271
personalizing your hacks and, 270
for specific targets, 267–69
ways to measure and monitor,
266–67
exercise:
supplements to prepare for
workouts, 131
work-hard philosophy and,
129–33
see also specific hacks and targets
eye, dominant, 185–86
eye movement desensitization and
reprocessing (EMDR), 232, 234

fats, 42, 51
monounsaturated, 65
polyunsaturated, 43, 65, 84
saturated, 42, 64–65
fiber, 66
insoluble (roughage), 66, 71–72
plants high in, 71–72
prebiotic, 92
soluble, 64, 66
fingernails, 65, 111, 112, 265
fish and seafood, 45, 47, 55–57, 109,
112
to avoid, 47, 56–57
varieties high in nutrients, 56
Fixx, Jim, 131
flavors, artificial, 57–58
flaxseeds, 71
fluorine, 64
FocusCalm neurofeedback system,
177
forgiveness, 219, 223, 226–28,
229–31, 258

40 Years of Zen, 22, 127, 176, 223, 229
free radicals, 41, 62, 88, 90, 93
free weights, 132, 134, 135–36
full occlusion, 168–69
fulvic minerals, 113, 209, 214
functional movement, 150–54, 256

gamma waves, 225
Gartenberg, Dan, 198–99
genetically modified foods (GMOs), 69–70
geranylgeraniol, 85
ginseng, 86–87, 91, 195, 202
glucaric acid, 102, 212
glucosamine, 40
glutathione, 111, 209, 212, 213, 232
glycine, 209, 212, 213
glycogen, 147, 148
glyphosate, 47–48, 69–70, 76
goat's milk, 53, 54
grains, 37, 39, 40–41, 48, 49, 55, 70–71, 99
grass-fed dairy products, 53–54, 56, 68–69, 83, 84, 106
grass-fed meat, 49, 67–68, 84, 94, 110
grass-fed whey protein, 51, 212
gratitude, 219, 226, 228, 229, 231
Gray, John, 145, 146
gray hair, 107, 108
Grossman, Dave, 143
Gundry, Steven, 39–40
gut, cleansing, 209, 214
gut microbiome, 46, 64, 66
    artificial additives and, 57
    glyphosate and, 47–48
    prebiotics, probiotics, and postbiotics and, 92–93

Hacks-Goals Matrix, 125–26
hair:
    gray, 107, 108
    stimulating growth of, 188–89
hair mineral analysis, 265
Hall effect, 170
hamstrings, 154
happiness, 260, 261–62
Hawkins, Jeff, 254
heart attacks, 62, 104, 131
heart disease, 84
heart rate, 16, 180, 199, 250
    cardio training and, 142–45, 146, 149

cardiovascular fitness and, 267
    measuring variability of (HRV), 236, 268
    potassium and fluctuations in, 105
heart rate variability, 199, 236, 241, 249, 268
heat and cold hacks, 126, 202–9
    controlled breathing with cold therapy, 203, 208–9
    cryotherapy, 203, 207, 210–11
    detoxification after, 210–11
    ice baths, 203, 207–8, 210
    sauna therapy, 202, 205–7, 210–11, 216
heat-shock proteins, 210
hemoencephalography (HEG), 176, 177–78
herbs, 72–73, 85–92
    adaptogenic, 90–91, 201–2; see also ashwagandha; ginseng; holy basil; rhodiola rosea
    cooking with, 86
    for resilience and recovery, 201–2
hero's journey, 59
high altitude, 166
    athletes training at, 163–64, 165
high-intensity interval training (HIIT), 120, 142, 145–46
    reduced exertion (REHIT), 120, 142, 145–46
Himalayan salt, 131
hip flexors, 154
histamines, 45–47, 92, 107
Hof, Wim, 164–65
    Wim Hof breathing and, 163, 165, 166, 208, 233
holotropic breathing, 219, 232, 233, 259
holy basil (Ocimum tenuiflorum), 88–89, 91, 195, 202
homeostasis, 16, 19, 23, 198, 202
homogenization, 53, 54
hormone replacement therapy, 111
hot and cold therapy, 213
Huberman, Andrew, 204
human growth hormone, 168, 196, 197–98
humic minerals, 112–13, 209, 214
humming, vocal vibration and, 157, 159
HUSO devices, 180
hyperoxic cardiovascular training, 142

hyperventilating, 164
hypothalamus, 187, 198
hypoxia, 121, 126, 162–66
    controlled, 163–64
    extreme, 163
    induced, with hypoxicator, 165–66
    local, in individual limb, 167

ice baths, 203, 207–8, 210
immune system, 45, 52, 62, 86, 202, 245
implants, 244–46
induced hypoxia, 165–66
inflammation, 46, 50, 52, 62, 196, 210
    lectins and, 39–41
    lessening, 74, 86, 87, 93, 94, 108, 199, 200, 201
    omega-6 fatty acids and, 41–43
    red-light therapy and, 199, 201
infrared light:
    in hacks for resilience and recovery, 199–201, 202, 205, 206–7
    hemoencephalography (HEG) and, 176, 177–78
    stimulating brain cells with, 185, 188–89
injuries:
    electrical stimulation therapy and, 170–71, 172
    wound healing and, 102, 157, 200, 206
insoluble fiber (roughage), 66, 71–72
insulin-like growth factor 1 (IGF-1), 168
interval training:
    high-intensity (HIIT), 120, 142, 145–46
    varying-intensity, 142, 144–45
iodine, 36, 64, 81, 109, 111, 113, 265
iontophoresis, 191–92
iron, 35, 36, 38, 48, 63, 68, 72, 81, 110, 200
isometrics, 134, 137

Jain, Naveen, 140
Jaquish, John, 138
Jobs, Steve, 12
Johnson, Bryan, 246–47
joints, 39, 40, 46, 47, 99, 109, 133, 135, 137, 171
    supplements and, 82, 95, 102, 110
junk light, removing, 185, 187, 197

kale, 38, 39, 70, 82, 117
keto diet, 103, 105, 108
kidneys, 48, 54, 105, 211, 214
    detox hacks for, 209, 210, 213, 215
    oxalic acid and, 38–39
kidney stones, 38, 78, 82
kindness, 1, 31, 61, 219, 226–28, 231, 235, 254, 258
Kotler, Steven, 139–40

lactic acid, 167
Lactobacillus plantarum, 92, 93
Lærdal Tunnel, 188
laser therapy, 185, 188–89
Lazarev, Nikolai Vasilyevich, 201–2
laziness principle, 2–7, 59, 130, 133
    body's autopilot function and, 2–3
    evaluating your hacking of, 264–65
    tapping power of, 11–30
    see also slope-of-the-curve biology
leafy greens, 38–39, 82
leaky gut syndrome, 39–40, 86
lectins, 38, 39–41, 50, 67, 70, 71
LED lights:
    blue-tinged, harm done by, 187
    brain blinking and, 185, 189–90
    modulated-light, 185, 188–89
    for red-light therapy, 201
leftover foods, 46, 47
legumes, 37–38, 39, 41, 47
libido enhancers, 86, 89
light hacks, 126, 184–90, 199–201
    LED brain blinking, 185, 189–90
    modulated-light LEDs and lasers, 185, 188–89
    reading under red light, 185, 186–87
    red-light therapy, 199, 201
    removing junk light, 185, 187, 197
    sound enhanced with light, 181, 184
    sun therapy, 199, 200
linoleic acid, 42
liver (food), 62, 77, 80, 81, 83, 107, 110, 112
liver (organ), 36, 46, 48, 54–55, 66, 69, 84, 88–89, 147
    detoxification of, 102, 209, 212–13
long-distance runners, hypoxia in training for, 163–64
longevity, 33, 66, 87, 89, 98, 150

as target for improvement, 118,
  120, 121, 122, 123, 126
  testing, 269
love, 17, 160, 223
L-theanine, 176, 195
L-tyrosine, 195
Lustig, Robert, 66
lymphatic system, 29, 158, 159, 214
  drainage from, 157, 162, 207, 209,
  211–12, 213

macrominerals, 101
Maffetone, Phil, 144
magnesium, 35, 48, 63, 71, 79, 97,
  101, 103–4, 113, 195, 200, 209,
  213
manganese, 35, 48, 64, 86, 110, 113,
  265
massage, 222
  lymphatic, 209, 211–12, 213
Masterjohn, Chris, 83
meals near bedtime, avoiding, 196
meat, 108, 112
  grass-fed, 49, 67–68, 84, 94, 110
  smoked, 47
MeatOS, 11–17, 20
meditation, 14, 22, 33, 76, 146, 179,
  184, 187, 197, 215, 219–20,
  224, 230, 237, 257, 259
melatonin, 187, 200
memory, improving, 86, 89, 91
mercury, 55, 111
mesominerals, 63
  see also iron
metabolism, 29, 36, 42, 48, 60, 62,
  64, 78, 93, 97, 210
  see also energy level and
  metabolism
methylated B vitamins, 156
microplastics, in fish and seafood,
  55–56
migraines, 103, 180
milk, tips on choosing, 52–54
minerals, 33–34, 35–36, 63–64, 96,
  97–113, 126, 131
  agricultural practices and, 33–34,
  98
  big, 63, 101–6; see also calcium;
  magnesium; phosphorus;
  potassium; sodium
  consuming right quantities and
  right proportions of, 101
  lab testing and, 265

mesominerals, 63; see also iron
multiple forms of, 100
trace, 35, 63–64, 106–12, 265;
  see also cobalt; fluorine; iodine;
  manganese; molybdenum;
  selenium; vanadium; zinc
  ultratrace, 36, 112–13
  vegan diets and, 38
  see also specific minerals
minitrampoline, rebounding on, 158,
  159
mitochondria, 24–26, 28, 43, 48, 75,
  77, 81, 84, 85, 110, 155, 196,
  210, 215, 253, 254
  body's energy system and, 24–25,
  41–42, 77, 110, 199
  breath work and, 164, 165, 166
  cardio training and, 144, 145, 147,
  148, 150
  energy supplements and, 93–94
  heat and cold and, 202, 203–4,
  206, 208, 210
  membranes of, 25, 41, 48, 94, 204,
  208
  reading under red light and, 186
  whole-body vibration and, 162
mitochondrial biogenesis, 25, 94, 148,
  150, 204
MitoSynergy, 107
mold toxins (mycotoxins), 54–55,
  209, 215
molybdenum, 36, 64, 111, 113, 265
monosodium glutamate (MSG),
  57–58
monounsaturated fats, 65
movement hacks, 150–54
  breathing and functional
  movement exercises, 150,
  152–54
  functional-movement
  consultation, 150, 151, 152,
  153, 154, 256
muscle, 15, 25, 26, 50, 65, 66, 68,
  74, 93, 94, 95, 102, 103, 121,
  125, 131
  blood flow restriction and, 167–70
  see also strength and muscle;
  strength training
muscle spasms, 103
Muse neurofeedback system, 177
Musk, Elon, 246
mycotoxins, 54–55, 209, 215
myelin, 191

NAD+ precursors, 156
Nautilus-style machines, 134, 136–37
nerve conduction study, 191
nervous system, 21, 123, 135, 175,
    202, 222
    circadian system and, 186–87
    electrical muscle stimulation (EMS)
        and, 134, 138–40, 190, 192–93
    resilience and, 123
    transcranial direct-current
        stimulation (tCDS) and, 190,
        191–92
    vagal nerve stimulation and,
        179–81
    vibrations and, 157, 158–59, 160
Nestor, James, 166
NeuFit, 140, 190, 193
Neuralink, 246
neurodegenerative diseases, 88, 91
neurofeedback, 175–79, 182, 225,
    235, 273
    expert-guided, 176, 178–79
    hemoencephalography (HEG),
        176, 177–78
    home devices for, 176, 177
neuro-fitness, see brain and neuro-
    fitness
neurological conditions, 46, 67, 89,
    91
neurological hacking, 175
neuroplasticity, increasing, 192
new normal, reaching for, 2, 273
niacin, 107
nickel, 112, 245
nightshade vegetables, 39, 40, 41
nitric oxide, 164, 167, 198, 200
NSAIDs, 105
nuts, 35, 38, 39, 44, 55

obesity, 26, 54
ochratoxin A, 54
oils, vegetable and seed, 42, 43, 44
olive oil, 43, 70
"Om," chanting, 158–59
omega-3 fatty acids, 65
omega-6 fatty acids, 41–44, 49–50,
    65, 69, 71, 84
oregano, 73
organic foods, 48, 49
orgasm, 232, 233
osteoblasts and osteoclasts, 157
osteoporosis, 35, 62, 78, 82, 97, 107
overweight, 26

oxalic acid, 38–39, 71, 99
oxaloacetate, 94, 156, 176
oxygen:
    blood flow restriction and,
        167–68
    controlled hypoxia and, 163–64
    electrical stimulation therapy and,
        171
    hyperventilating, then exhaling
        fully, 164–65
    protecting breathing during sleep
        and, 196, 197–98
    VO$_2$ max and, 120, 133, 147,
        149–50, 243, 267
oxygen hit, 149–50
oxytocin, 160, 180
oysters, 56, 107, 108, 110

pain relief, 103, 206
    functional movement expert
        consultation and, 152, 154
    vagal nerve stimulation and, 180
Panex ginseng, 86–87, 91
parasympathetic nervous system, 181,
    222
Parkinson's disease, 88
pasteurization, 52
peptides, 95
peroxisome proliferator-activated
    receptor gamma coactivator-1
    alpha (PGC-1α), 147, 204
personalized interventions, 239–42
phosphorus, 35, 38, 63, 79
phytase, 37, 71–72, 108
phytates, 35, 37, 47.50
phytic acid, 35–38, 40–41, 67, 70–71,
    73, 98–99, 108
pickled foods, 47
plank (yogic pose), 137, 157, 161
plant-based meat substitutes, 50, 51
plant-based protein powders, 50–51
plants, 69–73
    high in fiber, 71–72
    low in antinutrients, 70–71
    nutrient-rich, 72–73
pleasure foods, 73–74
polychlorinated biphenyls (PCBs), 56
polyphenols, 73
polyunsaturated fats, 43, 65, 84
polyvagal theory, 180, 222
Porges, Stephen, 179–80, 222
pork, 44, 45, 46, 50, 51
postbiotics, 92–93

potassium, 35, 63, 71, 72, 104–6,
    113, 213
    ace-K (acesulfame potassium), 58
    balance of sodium and, 101, 104–5
potassium bicarbonate, 106
potassium iodine, 109
Pranayama breathing, 164
prebiotics, 92, 93, 238
prefrontal cortex, 253–54
pressure cuffs, 167, 168–69
primal therapy, 221
probiotics, 92, 93, 97, 238
produce, 49, 69–73
propionic acid, 66
proprioceptors, overcoming resistance
    from, 134–41
protein, 51, 65–66
protein powder, 50–51, 129
protein powders, 50–51
psychedlics, 232, 234–35
psyllium, 71–72
pulsed electromagnetic field (PEMF)
    therapy, 171–73
pyrroloquinoline quinone (PQQ), 93,
    94, 156, 176

quantified self, 237–39
quinoa, 37

Rabin, David, 180
raw food diet, 99
reading under red light, 185, 186–87
reality, recognizing your lens on,
    253–54
rebounding, 158, 159, 211
recovery:
    after physiological task, 196–97
    spiritual, 225–31; see also spiritual
        resilience
    as target for improvement, 118–19,
        120, 121, 122, 123, 126, 129,
        194–95; see also resilience and
        recovery
red light, 256
    reading under, 185, 186–87
    for resilience and recovery, 197,
        199–201
    stimulating brain cells with, 185,
        188–89
reduced exertion high-intensity
    training (REHIT), 120, 142,
    146–48
regenerative agriculture, 98

relaxation:
    infrared sauna therapy and, 202,
        206–7
    vagal nerve stimulation and, 180
remove your friction, 27–28, 32–58
reset or restore mode, 160
resilience and recovery, 123, 194–216
    detox hacks for, 194–95, 209–16
    heat and cold hacks for, 202–9
    herbal hacks for, 201–2
    light hacks for, 199–201
    sleep hacks for, 196–99
    supplements to support, 195
    as target for improvement, 118–19,
        120, 121, 122, 123, 126, 129,
        194–95
    see also spiritual resilience
resistance bands, 134, 137–38, 140,
    154
Rhodiola rosea (aka arctic root or
    golden root), 89, 91, 195, 202
rhubarb, 38
rice, white vs. brown, 70
Robbins, Tony, 159
romantic attachment, 160, 180
rosemary, 73, 90, 91, 92
rye, 37, 71–72

Safe and Sound Protocol, 181,
    183–84
saffron, 86, 88, 91
sage, 89–90, 91
salad dressing, healthy, recipe for, 70
salmon, 56, 57
Salpeter, Garrett, 193
salt, 109
    bad reputation of, 104–5
    Himalayan, 131
    sea, 73, 106, 213
Sato, Yoshiaki, 169
saturated fats, 42, 64–65
sauna therapy, 202, 205–7, 210–11,
    216
    infrared, 202, 205, 206–7
seafood, see fish and seafood
seaweed, 109
Second Sight, 245
seed oils, 42, 44
seeds, 37, 38, 39, 44
selenium, 35, 36, 64, 111–12, 113,
    265
self-improvement, cycle of, 123–25
self-tracking, 239

Sensate devices, 160, 180–81
serotonin, 89, 200
sex, sexual function, 74, 96, 107, 259
  evaluating, 268–69
  pleasure and, 180, 232
  as spiritual practice, 233
  as target for improvement, 118,
    120, 121, 122, 123, 125, 126
sex hormones, 82, 85, 108, 109, 269
  see also specific hormones
Shapiro, Francine, 234
sheep's milk, 52, 53, 54
shellfish, see fish and seafood
shilajit, 113
shopping, 49, 187
Siberian ginseng, aka eleuthero, 87,
    91, 202
singing bowls, 159
single-photon emission computerized
    tomography (SPECT), 268
Sisson, Mark, 144, 145–46
sleep, 74, 118, 184, 187, 189, 196–99,
    200, 266
  deep and REM, 236, 269
  maintaining good sleep hygiene
    and, 196, 197
  monitoring, 196, 198–99
  protecting your breathing during,
    196, 197–98
  red-light therapy and, 199, 201
  reducing need for, 191
  sound devices and, 159
sleep apnea, 198
SleepSpace app, 196, 198–99
sleep tape, to hold lips together, 196,
    198
slope-of-the-curve biology, 2, 21–22,
    29, 30, 142, 143–44, 145–46,
    148, 156, 160, 163–65, 182,
    189, 194, 203, 207, 222, 224,
    226, 249, 266, 273
  see also laziness principle
smartphone, 245
smoked meat, 47
Søberg, Susanna, 210
social bonding, 160, 180
sodium, 35, 42, 63, 73, 100, 104–6,
    113, 213
  balance of potassium and, 101,
    104–5
soil-based organisms, 92, 93
soluble fiber, 64, 71
sonic feedback, 181, 182–83

"sound bath," 159, 160
sound therapy, 126, 181–84
Soviet astronauts, 190–91
soybeans and soy products, 38, 42,
    45, 46, 68
spices, 72–73
  cooking with, 86
  supplements, 85–92
spinach, 38, 39, 70, 106
spiritual guidance, 232–33
spiritual resilience, 219–36
  advancing from forgiveness and
    acceptance and, 230–31
  automatic kindness and forgiveness
    and, 226–28
  coming to terms with emotion and,
    220–23
  entering spiritual layer and,
    223–25
  spiritual growth hacks and,
    231–36
  spiritual reset and, 228–29
  spiritual stress and, 225–26
spot vibration, 158, 159–60
STAR Foundation, 221, 233
steady-state cardio, 142
Stephenson, Sean, 262
Strauss, Neil, 255
Stray-Gundersen, Jim, 169
strength and muscle:
  evaluating your outcome in,
    267–68
  functional movement and, 150–54,
    256
  as target for improvement, 118–19,
    120–21, 122, 123, 126, 129
  see also strength training
strength training, 134–41, 211
  AI-controlled machines, 134,
    140–41
  electrical muscle stimulation
    (EMS), 134, 138–40, 190,
    192–93
  with free weights, 132, 134,
    135–36
  isometrics, 134, 137
  Nautilus-style machines and
    weights with cables, 134,
    136–37
  overcoming resistance from your
    proprioceptors and, 134–41
  resistance bands, 134, 137–38,
    140, 154

supplements to prepare for, 131
work-hard philosophy and,
129–33
stress, 32, 266
adaptogens and, 201–2
detox systems hindered by, 215–16
evaluating your outcome in, 268
good vs. bad, 123
herb supplements for, 89, 90–91
reducing, 74, 143, 180, 182, 183;
see also resilience and recovery
spiritual, 225–26
as target for improvement, 118–19,
120, 121, 122, 123, 126, 129,
194–95
sucralose, 57, 58
sulfur, 35, 63
sun therapy, 199, 200
superoxide dismutase, 90
supplements, 75–96
amino acids and peptides, 94–95
energy, 93–94
herb and spice, 85–92
prebiotics, probiotics, and
postbiotics, 92–93
vitamin, 77–85
sweeteners, artificial, 57–58

tallow, 42, 65, 68
targets, 117–28
choosing, 117–19, 128, 266
cycle of self-improvement and,
123–25
Hacks-Goals Matrix and, 125–26
measuring and tracking progress
in, 127
see also specific targets
tech-guided breathing, 163, 165
teeth:
gritting or grinding, 198, 223
vitamins and minerals and, 38,
82, 99
temporomandibular joint (TMJ)
disorders, 180
testosterone, 62, 85, 125, 196,
197–98, 200, 269
Therastim, 139
threat, body's autopilot function and,
2–4, 23
thyme, 73
thyroid, 67, 80, 81, 108, 109, 111–12,
269
toxins:

biogenic, 54–55
chronic stress and, 215–16
glyphosate, 47–48, 69–70, 76
in one's own environment, 209,
214–15
see also detoxification
trace minerals, 35, 63–64, 106–12,
265
see also cobalt; fluorine; iodine;
manganese; molybdenum;
selenium; vanadium; zinc
transcranial alternating-current
stimulation (tACS), 191, 192
transcranial direct-current stimulation
(tCDS), 190, 191–92
transpersonal psychology, 219–20
trauma, 219, 222, 232, 234, 235–36,
251–52
trigeminal nerve, 175, 198
TrueDark, 187
turkey, 44, 50, 51
turmeric, 73, 86, 91

ultratrace minerals, 36, 112–13
Upgrade Labs, 29, 125, 127, 132,
140, 141, 148, 149, 150, 165,
171, 172, 206, 243–44, 273
urolithin A, 93

vagus nerve, 180
sound therapy with Safe and
Sound Protocol for, 181, 183–84
stimulation of, 179–81
vanadium, 36, 64, 112
varying-intensity interval training,
142, 144–45
vegan diets, 38, 60, 77, 80, 82, 99,
111, 264
vegetable oils, 42, 43, 44
vegetables, 39, 49, 92
cruciferous, 72, 109
mineral-deficient, 33–34
see also plants; specific vegetables
very-long-chain fatty acids (VLCFAs),
38
vibration hacks, 156–62
rebounding, 158, 159, 211
spot vibration, 158, 159–60
vagal nerve stimulation, 179–81
vocal vibration, 157, 158–59
whole-body vibration, 158,
160–62, 209, 211
vision exercises, 184, 185–86

vitamin A, 61–62, 77–78, 85, 265
    supplements, 80–81
vitamin D, 61–62, 75, 77–80, 85, 200,
    265
    supplements, 78–80
vitamin E, 61–63, 77–78, 85, 265
    supplements, 82–83
vitamin K, 61–62, 72, 77–82, 85, 265
    supplements, 81–82
VO₂ max, 120, 133, 147, 149–50,
    243, 267
vocal vibration, 157, 158–59
voltage-gated calcium channel,
    145–46

water, drinking, 213
    adding sea salt to, 73, 106, 213
water hyssop (Bacopa monnieri), 90,
    92, 176
wheat, 37, 39, 41, 47, 48
whey protein, 50, 51, 212
whole-body vibration, 158, 160–62,
    209, 211

Wim Hof breathing, 163, 165, 166,
    208, 233
work-hard philosophy, 129–33
workouts:
    supplements to prepare for, 131
    work-hard philosophy and,
        129–33
    see also specific targets and hacks
wound healing, 102, 157, 200, 206
wounds, mental, 228, 235, 252
Wozniak, Steve, 12
wrinkles, improving, 95, 200

yeast extract, 58
yoga, 152, 158
yogurt, 92, 106
"You do you," 251–62

zearalenone, 54
Zero Acre Farms, 43
zinc, 35, 38, 48, 56, 64, 68, 81, 100,
    107, 108, 113, 131, 200, 265
zone training, 142–44

ABOUT THE AUTHOR

DAVE ASPREY is the founder of Bulletproof, a four-time *New York Times* bestselling science author, and host of the top 100 podcast *The Human Upgrade*™ (formerly *Bulletproof Radio*), with more than 250 million downloads. Major news outlets, including the *Today Show*, CNN, *Wired*, *Good Morning America*, and Dr. Oz, call him the "Father of Biohacking" because he started the movement to take control of our own biology instead of just being "healthy." Dave hosts the world's largest and longest-running biohacking conference.

Over the last two decades, Dave has worked with world-renowned doctors, researchers, scientists, and global mavericks to uncover both new and ancient ways to increase longevity and mental and physical performance. He has personally spent nearly $2 million taking control of his biology, pushing the bounds of human possibility in the name of science and evolution, losing one hundred pounds, gaining a dozen IQ points, and taking eleven years off his biological age.

As a visionary and entrepreneur, Dave has founded multiple companies, including Bulletproof, the Upgrade Labs franchise, Danger Coffee, TrueDark/TrueLight, Homebiotic, and 40 Years of Zen, a revolutionary brain-upgrade clinic. In addition to his own corporate portfolio, Dave contributes as an investor and adviser to dozens of successful companies in the biohacking space. Today, he is focused on upgrading humanity in his role as CEO of Upgrade Labs, a franchised chain of Human Upgrade™ Centers. Dave also leads thousands of people enrolled in The Upgrade Collective, his online mentorship and membership group, where a community of people learn everything it takes to upgrade themselves at a physiological, psychological, and spiritual level.

Please head to DaveAsprey.com for information about the World of Dave Asprey.